NORTH CAROLINA AVIATRIX

VIOLA GENTRY

The Flying Cashier

JENNIFER BEAN BOWER

Foreword by Cris Takacs

THE History PRESS

Published by The History Press
Charleston, SC 29403
www.historypress.net

First published 2015

Manufactured in the United States

ISBN 978.1.60949.695.1

Library of Congress Control Number: 2014957214

Notice: The information in this book is true and complete to the best of our knowledge. It is offered without guarantee on the part of the author or The History Press. The author and The History Press disclaim all liability in connection with the use of this book.

In memory of Viola Estelle Gentry and in honor of Helen and Richard "Rich" Codling.

HER ACTIVITIES IN THE PROMOTION AND DEVELOPMENT OF AVIATION HAVE ENCOURAGED WOMEN THROUGHOUT THE WORLD TO PARTICIPATE IN THE FIELD OF AERONAUTICS.

—Tiny Broadwick Award plaque presented to Viola Gentry by the OX5 Club of America in 1965

Contents

Foreword, by Cris Takacs 7
Preface 9
Acknowledgements 11

1. East, West and Back Again: 1894–1924 15
2. Taking Flight: 1924–27 30
3. A Record Year: 1928 43
4. Triumph and Tragedy: 1929 57
5. Regaining Altitude: 1930–39 93
6. Changing Course: 1940–49 117
7. Flying Low: 1950–59 135
8. Flying High: 1960–69 143
9. Final Flight: 1970–88 165

Bibliography 181
Index 185
About the Author 192

Foreword

The International Women's Air & Space Museum (IWASM) was founded forty-five years ago by female pilots who knew that if something wasn't done, the history, voices and memories of the women from the early days of aviation would be lost. A collections manager's time is spent logging items and storing them in archival boxes on numbered shelves where they can be found in the future. Everything is interesting, but there are so many items that the collections manager doesn't get to linger over anything for too long. It is up to researchers to find the museum and bring the artifacts back to life.

Viola Gentry came into my field of view when a woman came to IWASM to do research on an airplane Viola had flown. The box of Viola's papers was brought out, and while the researcher went through them, she told me a bit about Viola Gentry. My concern at the time was the disorder the box was in and how the contents could be better organized. Viola's papers also contained many photos of other female pilots.

After the researcher left, work resumed on putting the papers and photos in better folders, labeling them and splitting the contents into two boxes. There was an extra copy of Viola Gentry's book *Hangar Flying* in the box, and the number was recorded in the database. Copies were made of the photos, and they were cross-referenced to other files of female pilots. Who was this Viola Gentry who seemed to have been in aviation for a great span of time? I made a note to read her book someday.

The book was not an autobiography as you would expect. When the weather is bad, pilots sit around in hangars, waiting for the sky to clear and telling tales of their flying experiences. This is known as "hangar flying." Viola Gentry's book is made up of these tales, and her own story is interspersed among tales of both the great and not so well-known pilots of the early days of flying. But what stories! After I finished the book, I opened the Viola Gentry boxes again to see if there was evidence that this woman had really done the things she said. The document evidence was there, as well as a little wallet of black silk that she had made to hold her various licenses and medical cards.

I began to notice more photos of Viola in the museum collections and came to know her handwriting on the back of them. She had been an early contributor to the museum archive. Anytime a photo of a group of women involved in the early days of aviation was brought out, I would peruse it and soon announce, "There she is," and whoever was looking with me would know that I was once again playing "Where's Viola," or "Spot the Viola."

Hangar Flying had come out in two forms, both privately published. IWASM has both the paperback and spiral bound versions. I eventually found a letter from Viola Gentry saying that IWASM could publish a better version with the photos she had gathered. That project has yet to be realized.

It was another copy of the book *Hangar Flying* that introduced me to Jennifer Bean Bower. She had found a copy and e-mailed IWASM requesting more information. Not having time to do anything myself with *Hangar Flying*, I was happy to help Jennifer with her project. A researcher has time (not more time, no one has "more time") to devote to the travel and research needed to flesh out the story of their subject. The IWASM files indicated that Viola Gentry had a niece. Jennifer found the niece and others who could tell her about Viola. Thanks to Jennifer Bean Bower, the museum is now the repository of Viola's memorabilia.

Charles Plank, in his 1942 book *Women with Wings*, referred to Viola Gentry as the selfless soul of aviation. The early days of aviation were full of colorful characters, and many of them had egos as big as the sky in which they flew. Viola Gentry was friends with all the aviation greats of her time, and she helped many attain fame. Now you will get to know this remarkable woman.

CRIS TAKACS
Collections Manager
International Women's Air & Space Museum, Cleveland, Ohio

Preface

As a native North Carolinian, I take pleasure in researching my state's history, heritage and culture. In particular, I seek out people, places and events that compose the lesser-known chapters of the state's history. While researching the origins of Maynard Field—North Carolina's first commercial airfield—in an effort to erect a historic marker near its site (the marker was unveiled in May 2008), I was introduced to Viola Estelle Gentry.

The State Archives of North Carolina and Dr. Thomas C. Parramore—who wrote the book *First to Fly: North Carolina and the Beginnings of Aviation*—documented Gentry as a pioneering female pilot from North Carolina. Because I was not familiar with her, I immediately wanted to learn more; but apart from a few articles and brief mentions in aviation history books, there was not much information available. For that reason, I mined historical documents to further my knowledge, and it wasn't long before the bits and pieces of Gentry's life began to take form.

From the moment Gentry decided to fly, she set a course for the sky and did not let anything—be it discrimination, financial hardship, misguided landings, crashes, physical limitations or tragedy—put an end to her dreams. Gentry's flight attempts and endeavors offer readers another chapter in the history of early aviation, one that reveals the struggles and achievements of a working-class woman with an unyielding will to fly. Her story is one of determination and inspiration; it is a narrative that deserves a place on the bookshelf of women's aviation history.

As Gentry's story is best told through her words, and those of her contemporaries, I attempted to write in a manner that would keep my voice distant. Through the use of extracted facts and quotes from primary source materials—particularly historical newspapers and Gentry's self-published book *Hangar Flying*—I feel this goal was achieved. *North Carolina Aviatrix Viola Gentry: The Flying Cashier* is not intended to reveal a complete history of women in aviation or aviation in general. It is simply meant to remember and honor the life of Viola Estelle Gentry, a native North Carolinian, a woman who proved—through her aviation ventures and accomplishments—that dreams *do* come true. In researching Gentry's life, I was encouraged to follow my own dreams. My hope is that you, the reader, will be inspired by her story as much as I have been and that your every dream takes flight.

Acknowledgements

This biography could not have been written without the assistance of many people. In documenting the trials and triumphs of Viola Estelle Gentry, it was necessary to spend countless hours researching historical documents. Those documents could not have been found and/or accessed without the assistance of staff at various museums, historical associations, libraries, genealogical societies and other repositories of aviation history. First and foremost, I must thank Cris Takacs at the International Women's Air & Space Museum in Cleveland, Ohio. Cris allowed me access to the museum's research files and provided many of the photographs seen within this book. She was also my constant "go-to" person whenever I had a women's aviation question. Thank you so much, Cris. Gentry's story could not have been told without you! Others who gave of their time, knowledge and energy to assist with the biography include Elizabeth Yuko and Helen Sammon, International Women's Air & Space Museum, Cleveland, Ohio; Laura Ohrenberg and Bonita Ades, the Ninety-Nines, Incorporated, International Organization of Women Pilots, Oklahoma City, Oklahoma; Denise Neil-Binion, 99s Museum of Women Pilots, Oklahoma City, Oklahoma; Chris Bentley, Tucson Chapter the Ninety-Nines, Incorporated, Tucson, Arizona; Christine Rousson, Fédération Aéronautique Internationale, Lausanne, Switzerland; Tina Ly, Tona Gates and Vicki Crocker, Federal Aviation Administration, Airman Certification Branch, Oklahoma City, Oklahoma; Julia Blum, Cradle of Aviation Museum, Garden City, New York; Elizabeth C. Borja, Kate Igoe and Danett Crespo, National Air and Space Museum,

Smithsonian Institution, Washington, D.C.; Rick Leisenring, Glenn Curtiss Museum, Hammondsport, New York; Debbie Seracini, San Diego Air and Space Museum, San Diego, California; Phoebe Bean and the Rhode Island Historical Society staff, Providence, Rhode Island; Rosser Lee Wayland, Danville Public Library and Virginia–North Carolina Genealogical Society, Danville, Virginia (thank you, Lee, for the tour of Danville and Gentry family home sites); Martinsville Public Library staff, Martinsville, Virginia; Sylvia Rowan and Christina Moretta, San Francisco History Center/San Francisco Public Library, San Francisco, California; Bonnie B. Coles, Library of Congress, Washington, D.C.; Dr. John Horner, Kansas City Public Library, Kansas City, Missouri; Mary Gomez, Gerald B. James Library, Rockingham Community College, Wentworth, North Carolina; Erik Huber, Queens Library, Jamaica, New York; Kathy Kauhl, Belleville Public Library and Information Center, Belleville, New Jersey; Tom Ankner, Newark Public Library, Newark, New Jersey; Stephanie Schmitz, Purdue University Archives, West Lafayette, Indiana; Dr. Thomas Allen, Special Collections, Eugene McDermott Library, University of Texas at Dallas, Richardson, Texas; N. Adam Watson and Kristen Gurciullo, State Archives of Florida, Tallahassee, Florida; Dawn Hugh, History of Miami, Miami, Florida; Maria David, *Charlotte Observer*, Charlotte, North Carolina; Stephen A. Kochersperger, United States Postal Service, Postal History Department, Washington, D.C.; Pat Ross, Bassett Historical Center, Bassett, Virginia; David Johnson, United States Army Ordnance Training and Heritage Center, Fort Lee, Virginia; Debra Long Hampton, artist, Winston-Salem, North Carolina; Don Buck and staff at the Western North Carolina Air Museum, Hendersonville, North Carolina (thank you, Don, for "use" of the Curtiss Robin and for the wonderful museum tour); and Diane Hein, Save the Biltmore Preservationists, Palm Harbor, Florida. To staff at various institutions that responded to my e-mails and phone calls, please know you are all appreciated.

In addition, many individuals, including Gentry family members and independent scholars, graciously allowed the use of personal and family files. Heartfelt thanks go to Helen and Richard "Rich" Codling, who welcomed me into their home and shared memories and memorabilia of their "auntie" with me. For that, I am forever grateful. Thank you also to Leon Gentry for showing me the locations of Gentry, North Carolina, and Gentry family landmarks; Lynn Myers for the photograph of George Henry Gee; Don Mabe for sharing his photograph of Nettie Walters Gentry's headstone;

Cindy Weigand and Allen Keller for information and photographs relating to John W. "Big Jack" Ashcraft; Linda Ann Marie Bertanzetti for sharing the photograph of Viola Gentry in the hospital; Harry Waterson for sharing photographs of a Buranelli medal and providing information on Felicity Buranelli and the Medal-of-the-Month Club; Gloria Conti Griffin for sharing remembrances and photographs of Viola Gentry at the Belleview-Biltmore Hotel in Florida; and to Angela Masson for providing documents, photographs and memories pertaining to Viola Gentry and the 1972 Powder Puff Derby.

For the opportunity to publish this book, I am indebted to The History Press, particularly Jessica Berzon, Adam Farrell, J. Banks Smither and Katie Stitely. Thank you for your assistance and belief in this project.

And thank you to my husband, Larry, my travel partner, research assistant, photographer and encourager. I would accomplish nothing without you.

1

East, West and Back Again

1894–1924

She was born before her time.
—Richard Gentry Hopper in Dwight Spark's article "Rockingham County's Famous Aviatrix," Madison Messenger, *December 3, 1980*

Above the sands of North Carolina's Kill Devil Hill, the first piloted, powered and heavier-than-air flying machine took flight. On that day—December 17, 1903—Orville and Wilbur Wright proved man could fly. In doing so, they secured North Carolina a prominent page in the annals of aviation history. Acclaimed for being "First in Flight," it should be known that North Carolina contributed more than a sandy knoll to the aviation timeline. Generations of North Carolinians, before and after the Wright brothers' historic flight, turned their eyes and minds skyward. Some invented flying machines, others helped win wars and many organized aviation-related businesses. A handful entertained the public with jaw-dropping aerial displays, while a few established their own flight records. Viola Estelle Gentry of Rockingham County, North Carolina, was one of the record setters. She was a trailblazing pilot—a celebrity in her own right—and this is her story.

Born on June 13, 1894, in the small community of Gentry, North Carolina, Viola Gentry spent a lifetime pursuing her aviation dreams. Traveling often in the companionship of hardship and tragedy, she weathered countless obstacles, the first of which came early in life.

In February 1900, Gentry and her younger sister, Thelma, were overcome by sorrow when their mother, Nettie Walters Gentry, died. Their father,

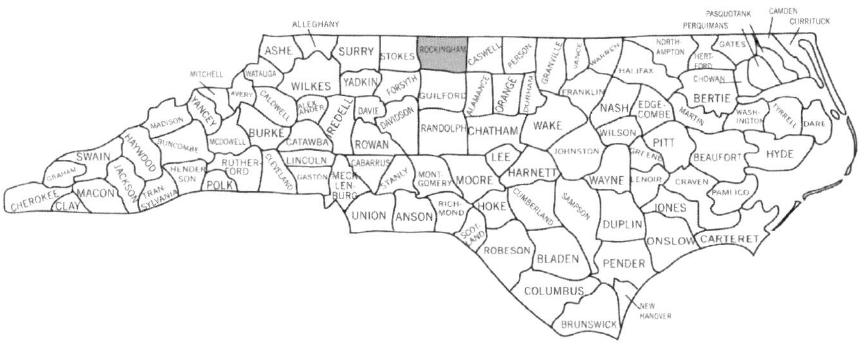
ALLEGHANY
ASHE
SURRY
STOKES
ROCKINGHAM
CASWELL
PERSON
GRANVILLE
VANCE
WARREN
NORTH-
AMPTON
HERT-
FORD
GATES
PASQUOTANK
PERQUIMANS
CAMDEN
CURRITUCK
MITCHELL
WATAUGA
AVERY
WILKES
YADKIN
FORSYTH
GUILFORD
ALAMANCE
ORANGE
DURHAM
HALIFAX
CHOWAN
FRANKLIN
BERTIE
YANCEY
CALDWELL
ALEX-
ANDER
IREDELL
DAVIE
DAVIDSON
NASH
EDGE-
COMBE
MARTIN
WASH-
INGTON
TYRRELL
DARE
MADISON
BURKE
CATAWBA
ROWAN
RANDOLPH
CHATHAM
WAKE
WILSON
PITT
HAYWOOD
BUNCOMBE
MCDOWELL
SWAIN
RUTHER-
FORD
CLEVELAND
LINCOLN
CABARRUS
LEE
JOHNSTON
GREENE
BEAUFORT
HYDE
GRAHAM
JACKSON
HENDER-
SON
GASTON
MECK-
LEN-
BURG
STANLY
MONT-
GOMERY
MOORE
HARNETT
WAYNE
LENOIR
CRAVEN
CHEROKEE
MACON
TRAN-
SYLVANIA
POLK
CUMBERLAND
SAMPSON
PAMLICO
CLAY
UNION
ANSON
RICH-
MOND
HOKE
JONES
DUPLIN
SCOT-
LAND
ONSLOW
CARTERET
ROBESON
BLADEN
PENDER
COLUMBUS
NEW
HANOVER
BRUNSWICK

Left: Nettie Walters Gentry, circa 1890. *Courtesy of Helen H. Codling. Right*: Samuel Garrett Gentry, 1901 or 1902. *Courtesy of Helen H. Codling.*

Opposite, top: The first piloted, powered and heavier-than-air flying machine took flight over North Carolina's Kill Devil Hill on December 17, 1903. Orville Wright is at the controls, while his brother Wilbur is seen running alongside the machine. *Courtesy of the Wright Brothers Negatives, Prints and Photographs Division, Library of Congress, Washington, D.C.*

Opposite, middle: Modern map of North Carolina with Rockingham County highlighted by Debra Long Hampton. *Courtesy of Debra Long Hampton.*

Opposite, bottom: The town of Gentry, North Carolina, was located south of Madison. This photograph, taken in 1979, is identified as the Gentry Post Office; however, oral history documents the building as an early Gentry family homesite. *Courtesy of the Rockingham Community College Foundation, Incorporated, Historical Collections, Gerald B. James Library.*

Samuel Garrett Gentry, who was a farmer at the time, struggled to work and care for his two young daughters. By June 1900, as evidenced in the United States Census, the motherless family had taken up residence with Samuel's father, Richard J. Gentry.

Samuel's younger siblings—Walter, Benson, Clyde and Anna—also resided in the home. Anna, who was twenty-seven years of age, became the girls' primary caregiver.

There is no question as to whether the living arrangement was agreeable to Gentry, as her recollections reveal a happy and loving environment. In her book *Hangar Flying* (the title is a term used to describe discussions had by pilots while sitting around a hangar), which is a collection of stories from early fliers—herself included—Gentry depicts her grandfather as a jovial storyteller. It seems his favorite tale, one he told with "great gusto," involved certain ancestors who fled to America under the charge of horse stealing. Gentry said her sister, Thelma, was not amused with his story, as she would have preferred pirates in the family tree. Their grandfather, upon hearing Thelma's grievance, "would just snort and retort, 'Pirate! Any fool could be a pirate. It took a smart man to steal a horse in England in those days—and not get caught!'"

Richard's animated behavior surely lessened the pain of the young girls' hearts, but the frivolities would not endure. In 1901, the structure of the family was disrupted when Anna married George W. Hopper. Although not documented, it can be assumed Anna departed her father's house shortly after her betrothal, which would have left Gentry and Thelma once again without a maternal figure. That was not the case for long.

By April 1902, Samuel and his two girls were living in Danville, Virginia, which is located approximately fifty miles northeast of the Gentry community. The reasons behind the move are unclear, but it might have been due to a job opportunity, a woman or both. In regard to Samuel's profession, *Richmond's Danville Directory 1902–1903* revealed he no longer worked as a farmer. Instead, he worked as a cigarette maker. Samuel married Maydie Blanche Price—a native of Virginia, who was

Mills and factories in Danville, Virginia, are documented in this 1930s postcard. *Courtesy of the author.*

fifteen years his junior—in May 1902. Afterward, the newly formed family resided in the household of Maydie's father, James R. Price. Two years later, they moved into a home of their own.

The years between 1900 and 1904 must have been difficult for Gentry and Thelma as they moved from one home—one family—to another. It is no surprise, therefore, that when the Wright brothers made history in 1903, Gentry (who was nine years old at the time) did not take notice of the event. She was a child who still mourned the loss of her mother, was obliged to create new relationships and pondered her own identity. The invention of flight held no relevance in her life—at least not at that moment.

Information regarding Samuel Gentry's family for the years 1905 through 1909 is scarce. However, in 1910, as documented in the United States Census, the family still lived in Danville. Samuel worked as a carpenter, and Maydie had given birth to two children—Lynwood Inez and James Forrest. Gentry, who was sixteen years old, labored as a roller at a cigar factory, while Thelma worked as a packer in a cigarette plant. For Gentry, the job no doubt proved tedious, as it did not coincide with her lively and adventurous spirit, a characteristic best described by her relatives.

In Dwight Spark's 1980 article "Rockingham County's Famous Aviatrix," which was published in the *Madison Messenger*, Richard Gentry Hopper, Anna's son, remembered Gentry as "a tomboy who used to go rabbit hunting with the menfolk."

Another description of Gentry's nature was provided by her niece Helen H. Codling (Thelma's daughter). In a written interview, Codling stated, "Before she got famous with stunt flying—she was known as that naughty Gentry girl."

Codling further declared that Samuel "got grey hair from [Gentry's] antics."

Gentry's personality was not suited for a job—or lifestyle—that required her to labor at the same task on a daily basis. She was a dreamer, a thrill seeker, a young lady who aspired for more than what Danville, Virginia, could offer—a fact proven in the article "Boys Follow the Circus/Five of Danville's Young Hopefuls Are Picked Up Here by the Police."

Published in the September 13, 1910 issue of the *Greensboro Telegram*, the article gave an account of two young boys, Claud and Al Harvey, who, after seeing the Robinson Circus in Danville, ran away from home. The boys followed the circus to Greensboro, North Carolina—an approximate forty-five-mile journey—where they hoped to acquire jobs. The Greensboro Police were notified, and the boys were apprehended and soon after "dispatched...back to their mother in Danville."

Two days later, as reported in the same article, the "police got wind that there was another bunch of young hopefuls [in town] from Danville."

This time, the group included three boys and one girl. The boys were Spot Graveley, Willie Warren and Otis Wells. The girl was Viola Gentry. When asked about their reason for wanting to join the circus, "They said the Harvey boys on returning to Danville told them that they could make $4.50 week and board working for the circus."

However, Gentry's and her young friends' dream of joining the circus came to a swift end, as "[Samuel Gentry] came in on No. 29 last night and carried the erring children back home."

That event, and the one to follow, is a good example of the antics that grayed Samuel's hair. In late 1910 or early to mid-1911, Gentry left home once again. For the cause of love, or perhaps in an effort to break free from the cigar factory, Gentry married George Henry Gee. It is not known whether Gentry's father knew of the marriage, but family recollection supports the notion he did not.

George Henry Gee, according to his World War I Draft Registration Card, was a tall, stout man with light hair and blue eyes. Photograph taken circa 1910. *Private collection.*

George, the son of Joshua and Elizabeth Gee, was two years older than Gentry. He lived in Danville and worked as a mill hand—presumably in one of the city's large textile mills. History does not reveal how the two met, where they were married or for how long, but a record on file at the Office of the Clerk of Circuit Court in Danville documents their divorce in 1911. The divorce was likely initiated by one or both sets of parents, which is evidenced by the fact that Gentry began living with an aunt and uncle in Jacksonville, Florida, in 1911. Perhaps the move was an attempt by Gentry's parents to keep her separated from George, or maybe they hoped a trip to Florida would satisfy her craving for travel and adventure. No matter their reason, the journey was a fortuitous one, as Florida was the location of Gentry's first flight.

During her stay in Jacksonville, Gentry toured the Florida Ostrich Farm with her aunt and uncle. A popular attraction, the Florida Ostrich Farm was the perfect outing for a young lady, as it offered a host of activities. There, tourists could view and learn about ostriches, cheer on their favorite bird during a harness race and shop for exquisite souvenirs. In addition, the farm also offered special programs and events. At the time of Gentry's visit, George A. Gray—a young pilot who had recently learned to fly at the Wright School of Aviation in Dayton, Ohio—was onsite and offering rides in his airplane. Eager to partake of the thrill, Gentry counted her money, found she had the amount needed, paid for the ride and took off. Unfortunately, she had failed to get permission from her aunt and uncle to do so. Therefore, the flight, and the consequences for taking it, became experiences she would never forget. As told by Douglas DeYounge Silver in his article "Cashier Weekdays—Aviator Sundays," which was written for the *Brooklyn Daily Eagle* on April 25, 1926, the "escapade…resulted in a sound spanking for its daring participant."

Silver further noted that Gentry, despite her punishment, became "profoundly absorbed in aviation" following her first ride in the sky. Yet her fascination with flight would not be revealed or nurtured for eight more years.

Postcard of the Florida Ostrich Farm, postmarked 1925. *Courtesy of the author.*

George A. Gray is seen center (bending forward) in this early twentieth-century photograph. *Courtesy of the State Archives of Florida, Florida Memory.*

Gentry's aunt and uncle, as demonstrated by their reaction to the flight, were not amused by her shenanigan. Perhaps for that reason, Gentry soon after returned to Danville. She did not, however, stay long.

As explained in *Hangar Flying*, Gentry was placed in the care of Mr. and Mrs. John Sears in 1912. John Sears was an electrical contractor and a friend of Gentry's father. The relationship between Gentry and this couple must have been friendly and/or long-standing, as she referred to them as "Uncle Johnnie" and "Auntie Sears." The purpose of the living arrangement is unclear but could have been a consequence of Gentry's behavior and/or relationship with George Gee. According to Gentry, Sears traveled extensively for his job, which allowed her to live "from a few weeks to several months in…thirty-eight different states…and…Canada." Without a doubt, Sears's travel requirements ensured Gentry's distance from both Danville and George Gee.

At the beginning of World War I, Gentry and the Searses were living in Bridgeport, Connecticut. The Connecticut residence must have been the Searses' permanent address, as they are recorded in Bridgeport city directories before and after the war.

While living in Bridgeport, Gentry attended one semester at a private school in Providence, Rhode Island. An adult at that time, it is not known what classes she took. Gentry remarked in *Hangar Flying*, "Uncle Johnnie took over my education and except for one term when I attended the Bronson-Hope School in Providence, Rhode Island, I had no other teacher."

As the war progressed, the United States found itself embroiled in the battle. American men and women, including Gentry and the Searses, were eager to support the war efforts. Mrs. Sears in particular became an avid promoter of Liberty Bond sales, as did Gentry. Together, the two sold a considerable amount of bonds, which Gentry proudly detailed in *Hangar Flying*:

> *Auntie Sears had become interested in promoting the sale of Liberty Bonds and worked out a plan which was approved by Ruth Pratt, who was Chairman of the Women's Liberty Bond Committee of the Second Federal Reserve District. To assist Auntie, I worked at meetings, rallies, etc., with three-year-old Louise Norman, who was a great-niece of Mrs. Sears. At a meeting, Louise and I would be on the stage and after the enthusiasm of the audience had risen to the point where someone would agree to buy a bond, Louise would salute the Flag and repeat the Pledge of Allegiance. Using this method, we sold over $75,000 in Liberty Bonds.*

Gentry's volunteer work with the Liberty Loan campaigns was also recorded in the October 12, 1918 issue of the *Bridgeport Telegram*. The headline read, "Poli's Audience Takes $10,000 in Bonds from Breen."

In order to sell bonds, volunteers, such as the Four Minute Men, women's groups and celebrities, would deliver patriotic speeches at concert halls and other public meeting venues. Their speeches were often accompanied by songs like George Cohen's "Over There," lantern slides and other patriotic displays—like little Louise Norman's flag salute and recitation of the Pledge of Allegiance. These patriotic endeavors would always end with a plea for funds. Such was the case at Poli's Theatre on the night of October 11, 1918. As documented in the article, Harry Breen, who was "one of Poli's headline performers…succeeded in selling [bonds] like 'hot cakes' until well over $10,000 were sold."

The article further reported, "Members of Miss Marguerite Beck's Minute Women's team together with other volunteers and a few men in uniforms assisted Mr. Breen in the sale of bonds. They went up and down the aisle filling out the blanks and receipts for the bonds as far as they were purchased."

At the end of the article, all those who had assisted Breen were named. The list included "Miss Viola Gentry."

Fueled by patriotic fervor, Gentry heeded the call of the Bridgeport District Ordnance Department and went to work as a clerk in its production division. Although the position was under the auspices of the United States Army, Gentry was a civilian employee. On November 11, 1918, World

War I came to an end. On April 1, 1919, her position with the Ordnance Department, like many other women's jobs across the nation, came to an end as well.

Gentry, who was twenty-four years of age and able to provide for herself, had to decide whether to remain in Bridgeport or return to Danville. In Danville, she would be reunited with her family and assured a position in a factory. Her ties to George Gee could not be reconnected, as he had remarried in 1915. George and his wife, Minnie Eanes, resided in Danville and, according to Helen H. Codling, were friends with Gentry's family. Perhaps for that reason, or the lack of job opportunities, Gentry did not return to Danville. She did not remain in Bridgeport either.

Following her discharge from the Ordnance Department, Gentry teamed up with the American Red Cross. The organization needed volunteers in San Francisco, California, to provide refreshments and encouragement to army recruits awaiting their overseas assignments. Ever a woman in search of a new experience or learning opportunity, Gentry no doubt viewed the chance to travel from one coast to another as the adventure of a lifetime. Little did she know, destiny had positioned her on a life-changing course.

By mid-1919, Gentry had said her goodbyes to Mr. and Mrs. Sears and left for California. Once there, she attended to the responsibilities assigned to her by the American Red Cross. Upon the successful departure of the army recruits, Gentry's services were no longer needed. Alone in a large city and without a job, Gentry found no cause to worry. She liked San Francisco, decided to stay and enrolled in a school for telephone operators. Upon completion of her courses, she became a switchboard operator at the Grand Hotel.

Gentry must have found San Francisco exhilarating. The Grand Hotel, as stated in an early twentieth-century brochure in the collection of the San Francisco History Center, San Francisco Public Library, was located "in the Heart of the Amusement and Big Department Store District." The brochure further noted, "The Grand Hotel is only a five-minutes' walk from San Francisco's main shopping district. Here also are the theatres."

Gentry rented a room on Turk Street, which was located within a block of the Grand Hotel. She not only worked within the city's main thoroughfare but resided there, too. A smart choice for work and housing, the location ensured that all of Gentry's needs—the post office, bank, restaurants, clothing, etc.—could be met within a short walk.

The constant hustle and bustle of the city provided an endless array of entertainment and new experiences for Gentry; but although the city itself was vibrant, her job was not. Much like her former position in the cigar

Early twentieth-century postcard of the Grand Hotel in San Francisco, California. *Courtesy of the San Francisco History Center, San Francisco Public Library.*

factory, Gentry was restricted to a singular location for long periods of time where she performed the same actions again and again. The position must have proved monotonous, particularly her conversations, which, in all probability, consisted of "room number?" and "one moment please."

Nevertheless, on an eventful day in July 1920, Gentry witnessed a daring act, a feat that reawakened her passion for flight. She spoke of the occurrence often and described it at length in *Hangar Flying*:

> *When I reported for work one morning, the hotel was buzzing with excitement. Men, they said, were busy putting chicken wire around the top of the St. Francis Hotel so that a plane could be safely landed on the roof. Lt. Ormer Locklear, an ex-army pilot, had turned to stunt flying and had been doing some rather spectacular things in the air. It seems that on this day he was to land a plane on the roof of the St. Francis Hotel as part of a stunt for Universal Pictures* [the stunt was for a Fox Film Corporation production] *and everyone was speculating on whether or not he would succeed. During my lunch hour I went over to the St. Francis Hotel to see what was going on.*

Ormer L. Locklear, who held the rank of second lieutenant during World War I, was a talented stunt pilot who performed aerial feats for Hollywood

Left: Ormer L. Locklear, 1920. *Courtesy of the National Air and Space Museum, Smithsonian Institution (SI 85–12330).*

Below: Postcard of the St. Francis Hotel in San Francisco, California, postmarked 1921. *Courtesy of the author.*

movies. On Monday, July 12, 1920, Locklear arrived in San Francisco and registered at the St. Francis Hotel. According to the article "Several Movie Stars Are Visiting in City," which appeared in the July 12, 1920 issue of the *San Francisco Chronicle*, Locklear and other movie stars were in town "to make scenes for *The Skywayman*, a new production starring the daring aviator."

When Gentry arrived at the St. Francis Hotel, she saw Locklear flying strategically around the building. The crowd around her gasped and shouted as the distance between airplane and roof shortened. Within minutes, Locklear landed in perfect form and was heralded with a chorus of cheers. For Gentry, the spectacle was more than a few moments of breathless entertainment; it was a life changer.

H. Glenn Buffington, who interviewed Gentry for his 1968 article "Flying Life of Viola Gentry," documented her as saying, "That did it!" upon seeing Locklear's aerial display. In *Hangar Flying*, Gentry remarked that Locklear's stunt "seemed very easy," and "if a man could do it, certainly a woman could. All she had to do was learn to fly."

Gentry's enthusiasm for aviation, an interest buried since her first flight in Florida, had been resurrected and would not be suppressed again. Gentry returned to work that afternoon, but her thoughts were fixated in the clouds. Ready and eager to learn everything about airplanes and flying, Gentry began attending lectures on flight, reading books on the subject, asking questions of any pilot who would speak with her and putting money aside for her first flying lesson. Her appetite for all things flight-related had become insatiable.

After saving enough money for a lesson, Gentry arranged to meet a flight instructor at Crissy Field in San Francisco. There is no doubt that she was ecstatic as she climbed into the airplane, placed her hands on the controls and awaited instructions. However, her excitement was dampened, albeit not extinguished, by the words of her instructor, Robert "Bob" Fowler—the first man to fly from the West to East Coast. He said, "A woman should NOT fly, but should stay home, get married and raise a family."

His statement, which was recorded in *Hangar Flying*, was not an uncommon opinion at that time. According to Kathleen L. Brooks Pazmany, in her book *United States Women in Aviation 1919–1929*, "The attitude that flying was socially inappropriate and even physically impossible for women was common."

Gentry, of course, did not agree with Fowler's sentiments, but as she only had enough money for one lesson, she was "grounded anyway."

In September 1924, Gentry quit her position at the Grand Hotel and left San Francisco. She was positive New York would provide

Crissy Field in San Francisco, California, circa 1930. *Courtesy of the author.*

Robert "Bob" Fowler, circa 1910. *Courtesy of the Bain Collection, Prints and Photographs Division, Library of Congress, Washington, D.C.*

better opportunities for women interested in flight. Before leaving the state of California, Gentry traveled to Santa Monica to see the Round-the-World Fliers—United States Army pilots who had successfully circumnavigated the globe—who were making the final stops of their around-the-world journey.

It is interesting Gentry chose to see the fliers at Clover Field in Santa Monica, which was nearly four hundred miles south of her residence, instead of at Crissy Field in San Francisco. The decision to do so must have related to her travel plans, as she had made arrangements to leave California on September 24. The Round-the-World-Fliers were scheduled to arrive in Santa Monica on September 23 but would not reach San Francisco until September 25. Therefore, her only chance to see them was in Santa Monica.

At Clover Field, Gentry had the fortuity to speak with Lieutenants Lowell Smith and Leigh Wade, two of the Round-the-World Fliers. She recalled the men's encouraging words when she wrote of the event in *Hangar Flying*:

> *It was a privilege to meet them and I remember with pleasure, the kindness of Lt. Lowell Smith and Lt. Leigh Wade who took the time to talk to a woman who wanted to fly. They told me they believed a woman could do anything she wanted to in the air and that the women flyers they knew were as good or better than the average man flyer—if I wanted to fly with all my heart and soul and loved everything connected with flying, I should keep on and not let anything discourage me—there would be a place for me wherever I might be—and I should just take a look at Marvel Crosson* [Crosson, the first woman to obtain a pilot's license in Alaska, was at Clover Field that day] *and know that somewhere, sometime, I too, would fly.*

The next day, Gentry, fueled by the words of the famous fliers, headed east.

2

Taking Flight

1924–27

When I began to fly…my family recoiled in horror.
—Viola Gentry, Hangar Flying, *1975*

Gentry arrived in New York with a big dream and a small amount of money. She had made arrangements to stay with a friend until her plan—a creative scheme to learn how to fly at little or no cost—could be put into place. Should the idea fail, Gentry would get a job, find a place to live and take flying lessons in her off hours.

"If you wanted to [learn to] fly," said Gentry in *Hangar Flying*, "you got yourself into a plane and flew."

That was easier said than done. To get into an airplane, Gentry had to rent one, and to fly it, she needed a flight instructor—both of which cost considerable amounts of money. An average student required ten to twenty hours of flight instruction prior to their first solo flight, and the combined fee for airplane rental and flying lessons, as reported in *Hangar Flying*, was sixty dollars an hour.

The United States Internal Revenue Service in its booklet *Statistics of Income from Returns of Net Income for 1924*, reported a majority of Americans as earning less than $2,000 a year. For Gentry, the cost of one flying lesson equaled approximately three weeks' wages. It is easy to understand, based on that fact, why she was eager to find—or create—little- or no-cost opportunities to get in the air. Tom Parramore, in his article "Viola Gentry," best explained Gentry's financial dilemma when he stated, "She was among the few blue-collar participants in a society of the gilded elite."

When Gentry learned Philadelphia, Pennsylvania, would host a world's fair in 1926—the Sesquicentennial International Exposition—to celebrate the 150th anniversary of the signing of the Declaration of Independence, she saw an opportunity. Although the fair would not take place for two years, city leaders had already begun planning, preparing for and promoting it. Gentry came up with an idea she believed would create substantial publicity for the celebration and allow her to learn to fly, free of charge, at the same time.

Before presenting the idea, Gentry wanted to ensure she could answer any questions the event organizers might put before her. To do so, she met with Charles S. "Casey" Jones, manager of Curtiss Field in Valley Stream, Long Island, New York. Gentry told Jones she wanted to persuade the city leaders of Philadelphia to let her fly, with a pilot, to each of the fifty state capitals, where she would present the governor a personal invitation to attend the Sesquicentennial International Exposition. "When I told him what I had in mind," Gentry said in *Hangar Flying*, "he looked at me as if he thought I was ready for the men in white coats to take me away. I had to convince him that I was never more serious in my life."

And convince him she did. Jones and his assistant, Mazel M. Merrill, were impressed by Gentry's fortitude and gave of their time and knowledge to assist her. The men worked with Gentry to determine the cost of the trip and the routes to be taken. Along with other pertinent facts, they showed Gentry how to read a Rand McNally road map and chart a course. After a month of discussions, she requested a meeting with the mayor of Philadelphia. Her appeal was received and an appointment scheduled.

On the morning of her meeting, Gentry entered a room in which the mayor, a representative of the governor of Pennsylvania, ten other men and a stenographer were waiting to hear her idea. Gentry, with notes in hand, explained to the committee every detail of her plan—how much it would cost, how long it would take and why it would be profitable. She waited with great hope to hear their responses, to answer their questions; instead, she came under attack. When writing of the event in *Hangar Flying*, Gentry said:

> *They wanted to know why I thought it was becoming for a lady to fly with a man pilot to all these states to present the invitations…couldn't a man do it just as well alone…why I thought I should be the lady to do that sort of thing…why couldn't they select someone else if they chose…and why did I think I had a monopoly on the idea. I never heard so many "whys" and "why-nots" and so many reasons why a woman should presume to do*

> *anything of the sort and began to wonder why they thought a woman should ever presume to be born.*

Gentry listened, just as she had in San Francisco, to the discriminatory remarks but was neither silenced nor deterred by them. In response to their statements, Gentry did not address women's rights or whether it was appropriate for them to fly. She spoke only of her plan's ability to generate increased exposure and revenue for the event. Gentry told the committee she "would get publicity for the Sesquicentennial which could never be paid for."

She further stated that the "front pages of a newspaper are never available for advertising, but the story of a fledgling pilot going from capital to capital to deliver an invitation would make the front pages of newspapers everywhere."

Gentry's statements, however, fell on deaf ears. The committee's objections to her plan rested on one simple premise: "that it was unbecoming for a lady to fly."

The city leaders thanked Gentry for submitting the idea and said they would mail her a letter regarding their final decision. Gentry returned to New York disappointed but eager to receive the letter. She knew what it would say, but until she saw the answer in writing, there was a chance the committee could say yes.

When the letter from Philadelphia arrived, Gentry read what she already knew: the committee was not interested in her plan. If there was any consolation to the committee's disregard for Gentry's idea, it came in the form of failure. In regard to the Sesquicentennial International Exposition, she concluded:

> *And—as a final note on the Sesquicentennial, I might add that the affair would wind up with a loss reported to be over five million dollars. I can't say that this loss would have been wiped* [out] *or diminished if my plans had been used—they had so much to contend with…But I like to think that my plan would have given them the added publicity which they badly needed—at a moderate cost.*

One might suppose Gentry made those statements as a result of ill feelings toward the event, but the article "Too Much to See," which was published in the October 18, 1926 issue of the *Iowa City Press-Citizen*, confirmed her sentiments. It stated, "The Sesquicentennial International exposition at Philadelphia is a financial failure…since the opening, only 5,000,000 have attended. Attendance of 25,000,000 was expected."

Perhaps the outcome would have been different had the committee put Gentry's idea in place, but that fact will never be known. It is known,

Postcard of the Sesquicentennial International Exposition, Philadelphia, Pennsylvania, 1926. *Courtesy of the author.*

however, that Gentry did exactly what she said she would do if her plan failed: she became a cashier at Madame Barna's French Restaurant, rented a small apartment in Brooklyn and secured a second job as a hat check girl in order to save money for flying lessons.

Gentry's shift at Madame Barna's began daily at 11:00 a.m. The schedule was perfect because it allowed her to spend several hours each morning at Curtiss Field. On days she did not have enough money to take flying lessons, she spoke to and asked questions of pilots, all of whom were helpful and responsive. To learn more about airplanes, Gentry read books and volunteered her assistance to anyone at the field who would take it. She was determined to be a pilot and understand the machines she would fly. A permanent fixture at Curtiss Field, Gentry gained the respect of all who came to know her.

For the remainder of 1924 and throughout 1925, Gentry continued to study the technical aspects of aviation and took flying lessons when she could afford them. George Vecsey, who interviewed Gentry in 1977 for his and George Dade's book *Getting Off the Ground*, documented her flight instructor as Roger Q. Williams, a pilot from Brooklyn who established the Roger Q. Williams School of Aeronautics and set a transatlantic flight record. Vecsey also recorded the month and year in which Gentry first soloed: September 1925.

Opening day at Curtiss Field, May 15, 1921. *Courtesy of the Curtiss-Wright Corporation Records (CW8G-T-1868), National Air and Space Museum, Smithsonian Institution.*

Gentry had at last earned her wings, and from that day forward, her life would never be the same. She had studied, saved her money and conquered the sky, but it was not enough. Shortly after her solo—which, according to Gentry in a 1971 interview at the OX5 Aviation Pioneers Annual Meeting in Phoenix, Arizona, was in a Curtiss JN-4 ("Jenny" biplane) with an OX5 engine—she applied for and received an aviator's certificate from the Fédération Aéronautique Internationale (FAI) and National Aeronautic Association of the United States (NAA). The NAA is the official United States representative of the FAI. The certificate was important because it gave her credibility as a pilot and would be necessary for her to carry out her plans, which included setting and breaking aviation records.

It is interesting to note that Gentry's 1926 aviator certificate revealed the year of her birth as 1900. Gentry was born in 1894 but, after her solo, gave 1900 as the year of her birth. Although it is not known why she did this, it might have been in response to the young ages of other women who were flying at that time.

Gentry desired a career in aviation, but more importantly, she wanted to make a name for herself and prove women were skilled pilots. Gentry spoke of this moment in her life in *Hangar Flying*. She said, "Now my brain began to work overtime trying to think of something which would help publicize aviation as a career for women. Sure—we could fly, but how could we make the public understand that we were not only good pilots but could do all the things in the air which were being done by the best men pilots?"

FÉDÉRATION AÉRONAUTIQUE INTERNATIONALE

NATIONAL AERONAUTIC ASSOCIATION OF U. S. A. INC.

Certificate No. 6530

The above named Association, recognized by the Fédération Aéronautique Internationale, as the governing authority for the United States of America, certifies that

VIOLA GENTRY

born 13 day of JUNE 1900

having fulfilled all the conditions required by the Fédération Aéronautique Internationale, for an Aviator Pilot is hereby brevetted as such.

Dated SEPT 16 1926

CONTEST COMMITTEE

Orville Wright Chairman

Executive Vice-Chairman

Signature of pilot: Viola Gentry

Viola Gentry's Fédération Aéronautique Internationale/National Aeronautic Association of the United States aviator's certificate, dated September 16, 1926. *Courtesy of Helen H. Codling.*

Throughout 1925 and early 1926, Gentry worked various jobs, rented airplanes when she could afford it, logged flight time and befriended many pilots (men and women) who left their mark on aviation history. She also continued plotting and planning ways to publicize her own flying abilities. As Gentry said in *Hangar Flying*, "I worked hard, got into the air every time I could—and waited."

Gentry wanted her first flying stunt to leave an impression on New York, so she decided to fly underneath the Brooklyn Bridge, a bridge she often crossed by foot. During one of her near daily walks, Gentry met a patrolman named Benjamin Foster who she told of her intentions. Foster did not think the plan was a good one, yet he informed her who to contact for permission to do so. Gentry mailed a letter to the appropriate city department but did not receive a response. Nevertheless, she was determined to make the flight.

On March 14, 1926, Gentry rented a Curtiss Oriole, a three-seat biplane, at Curtiss Field and took off. She was not alone, however, because Curtiss officials knew of her intentions and required her to make the flight with

Arthur "Art" Caperton, a veteran Curtiss pilot. Gentry provided an account of the death-defying stunt in *Hangar Flying*:

> *We circled the Statue of Liberty and then started up the river. We zipped under Brooklyn Bridge and were caught in vicious air currents. As we cleared the bridge we were caught in a side wind from the west—rose high over the river for a minute—then ducked again for Manhattan Bridge, which was just ahead. We cleared that safely and then flew off for Curtiss Field. I—a fledgling pilot with little solo time—had flown underneath both Brooklyn and Manhattan Bridges. As we landed, we were met by several of the reporters from New York newspapers, and I was so pleased when I heard one of them talking to Art Caperton, who said: "It was quite a ride, but I'm glad it's over. Believe me, I was mighty glad to get out from under those bridges. The air was sure choppy. And that girl is a mighty good pilot all right. I never worried about her."*
>
> *It really had not occurred to me to be frightened, but I sort of shivered later when I thought that if I had made a mistake, we would have gone under in our heavy flying clothes* [and] *we never could have been fished out of the river.*

Gentry took a chance, showcased her flying abilities and succeeded in making a name for herself. When she returned to Curtiss Field, she found herself surrounded by reporters. The next day, her name—and, in some instances, her photograph—was included in newspapers across the country. In New York, the *Brooklyn Daily Eagle* ran the headline, "Girl Flier Dives Under Bridges of East River for Thrill After Hiring Plane for Daring Jaunt."

The article provided a brief, albeit incorrect, account of Gentry's personal life and flying record, as well as an interview with Art Caperton, who said, "I do not know what her purpose was in making the flight unless it was just that she wanted a thrill. Her stunt is not a record-breaker, however, as many planes have been piloted under those bridges. As far as I know, she is the first woman to do it."

Although Gentry has not been recognized as the first woman to fly underneath the Brooklyn and Manhattan Bridges, subsequent newspaper and magazine articles, and Gentry herself, claimed this fact to be true. The article "Cashier Weekdays—Aviator Sundays," which was written by Douglas DeYounge Silver for the April 25, 1926 issue of the *Brooklyn Daily Eagle*, declared Gentry's flight "the first safe and successful trip of a woman-piloted airplane underneath Brooklyn Bridge."

Early twentieth-century postcard of the Brooklyn and Manhattan Bridges in New York. *Courtesy of the author.*

Underneath the Brooklyn and Manhattan Bridges, circa 1909. *Courtesy of the Detroit Publishing Company Photograph Collection, Prints and Photographs Division, Library of Congress, Washington, D.C.*

Viola Gentry was photographed on March 14, 1926, after her flight underneath the Brooklyn and Manhattan Bridges. *Courtesy of the National Air and Space Museum, Smithsonian Institution (SI 80–451).*

Other publications also shared news of Gentry's bridge stunt with their readers. On March 15, 1926, the *Brooklyn Standard Union* included a photograph of Gentry's airplane as it was about to fly underneath the Brooklyn Bridge. The headline read "Brooklyn Girl Takes Flyer Under Bridge," while the caption beneath the photograph stated, "Just to show that one can take the air under Brooklyn Bridge as well as over it Viola Gentry, aviatrix, sent her aircraft whizzing under the big span."

The March 15 issue of the *Ogden Standard-Examiner*, in an Associated Press column titled "Flashes of Life," reported, "New York.—Miss Viola Gentry, cashier in a restaurant, has piloted a plane beneath the Brooklyn and Manhattan bridges, 133 and 135 feet, respectively, above the water line. A male pilot was with her. She says she has a pilot's certificate herself and needed a thrill that her regular job does not provide."

A section of the *Bee*—a Danville, Virginia newspaper—titled "News of the World Told in Pictures" included a photograph of Gentry on March 19 with the caption, "Daredevil—Miss Viola Gentry, cashier in a New York restaurant, got tired of the monotony, borrowed an airplane and flew through the arch of Brooklyn Bridge."

If Gentry's family had been unaware of her flying activities and interests, they were likely shocked to see her face within the pages of their local newspaper. In reality, they might have found the news appalling. Robert Carroll, who was Gentry's second cousin, wrote a column for the *Danbury Reporter* titled "Old, Odd & Other Stuff." In the December 21, 1988 issue of the newspaper, he wrote about Gentry's flight. In the article, Carroll said, "To the very prim Gentry family Viola's activities were so unladylike that they felt she had disgraced the family. It was a 'hush hush' affair and her family connections were 'a skeleton in the closet.'"

There is no doubt Gentry's flying endeavors alarmed her family, particularly her father, Samuel, and stepmother, Maydie. As Gentry said in *Hangar Flying*, "For a woman to pilot an airplane is taken quite as a matter of course now, but it was simply unheard of when I began to fly and my family recoiled in horror."

Days after Gentry completed her bridge stunt, newspapers were still covering it. A page in the March 16, 1926 issue of the *Philadelphia Enquirer*—a section titled the same as in the *Bee*—"News of the World Told in Pictures," showed a close-up image of Gentry's airplane passing underneath the Brooklyn Bridge. The caption read, "'Does Stunt Flying on "Days Off"'/ During the week Viola Gentry is a restaurant cashier in New York, but last Sunday, she stepped from this role and, in a plane, flew under Brooklyn bridge, where dangerous air currents have forced expert fliers to evade the same stunt."

Douglas DeYounge Silver, in his full-page article "Cashier Weekdays—Aviator Sundays," recounted an interview he had with Gentry inside her Brooklyn apartment. Most of his questions related to the bridge stunt—why, when and how she decided to do it—while others revealed personal details of Gentry's life. In regard to Gentry's appearance and character, Silver noted she was "a pretty, brown-haired and demure restaurant cashier" who "was most cordial" and had a "spick and span little apartment."

He then provided an interesting commentary regarding her thoughts on automobiles. Silver reported, "Miss Gentry doesn't like automobiles, can't drive one because she gets all mixed up with the gearshift, and anyway she feels 100 percent safer in an airplane—so naturally she prefers this method of locomotion."

At the end of the article, Silver gave a summary of Gentry's interests. He also told readers about a ritual performed by Gentry at the end of each day. His last two paragraphs read:

> *An unusual girl. A girl who gets a great kick out of looping the loop, doesn't class it as a stunt, and who gets fun from making most of her own clothes and from her restaurant job besides. She likes being a cashier because she likes people and enjoys studying them. For amusement she likes any type of show except ornate musical comedies. Her favorite authors are H.G. Wells, Maxim Gorky, Upton Sinclair, and Zane Grey.*
>
> *Every night she has one important task. Viola Gentry's mother died many years ago, so long ago that she can hardly be remembered by her daughter in New York. Nevertheless, each night after she comes back from work this daughter writes a letter to her mother—just as though she were alive today. These letters, which are kept in a large book, form the real story of the life of one of America's most outstanding young women.*

Two images accompanied the article; one was a photograph of Gentry, and the other was a drawing of a cash register with airplane wings. On top of the cash register was a female pilot, and in the distance was the Brooklyn Bridge. Following the publication of the article, thanks largely (one can assume) to that drawing, Gentry became known as the "Flying Cashier."

Gentry was not the only pilot to have an informative term attached to his or her name. For example, Lieutenant Belvin W. Maynard, an ordained minister who won the Transcontinental Air Race of 1919, was called the

"Flying Parson"; Elinor Smith, who became a licensed pilot at the age of sixteen, acquired the nickname the "Flying Flapper"; and Amelia Earhart, whose physique somewhat resembled Charles Lindbergh, was often referred to as "Lady Lindy."

Nevertheless, Gentry did not like her newly acquired moniker. James Kilgallen, who interviewed Gentry for his article "Girl Pawned Gems to Become Flyer and Break Record," which was published in the December 24, 1928 issue of the *Kane Republican*, reported, "Miss Gentry dislikes being termed the "flying cashier" because she worked as a cashier in several New York restaurants to earn expenses incurred while flying."

Gentry, however, didn't mind being called "the pest" by pilots she knew. According to Kilgallen, Gentry said, "[I] was so intensely interested [in aviation] I became known as 'the pest.' I was always around the flying machines and the aviators. They couldn't get rid of me. Finally it got so when I didn't show up at the hangar or at the restaurant the good pilots would ask, 'Where's the pest?'"

The media latched on to the term "Flying Cashier" and consistently used it alongside Gentry's name. However, neither her title nor the notoriety of flying underneath the Brooklyn and Manhattan Bridges provided any prosperity. If anything, the bridge stunt brought additional debt. The *Brooklyn Daily Eagle* confirmed this fact in the March 15, 1926 article "Girl Flier Dives Under Bridges of East River for Thrill After Hiring Plane for Daring Jaunt." In it, the author stated, "An official of the Curtiss company was asked how much it cost Miss Gentry to hire the plane. He said…that because of the nature of the young woman's flight, she had to be on a basis of $100 an hour. She was out a little less than an hour."

Months later, New Yorkers were still interested in Gentry and her flying escapades. In August 1926, Captain René Fonck, a French flying ace, was the featured guest on the New York radio station WRNY. Along with Captain Fonck, other notable pilots were invited to speak during the program. According to the article "On the Radio Last Night," by L-S-N-R, which was published in the *Brooklyn Daily Eagle* on August 28, 1926, Fonck addressed listeners in his "native tongue," which was later "read…in English." After Fonck's speech, "messages from President Coolidge, Governor Smith and Mayor Walker were read. The French consul-general said a few words, as did Col. H.G. Hartney, of the Argonauts…Other speakers were Lieutenant Snoddy…Viola Gentry, who recently flew under the East River bridges and who proposes to do some bigger stunts; and Walter Weltman, the Artic explorer."

Viola Gentry's Fédération Aéronautique Internationale/National Aeronautic Association of the United States annual sporting license, issued in 1927. *Courtesy of the International Women's Air & Space Museum, Cleveland, Ohio.*

For the duration of 1926 and throughout 1927, Gentry worked multiple jobs, frequented Curtiss Field and maintained her sporting license with the FAI/NAA. When the Air Commerce Act of 1926 went into effect—an act in which the United States government began regulating all aspects of aviation and issuing pilots' licenses (the first license was issued on April 6, 1927, to William P. MacCracken Jr., assistant secretary of commerce for aeronautics)—Gentry at once took the appropriate steps to secure her license. In September 1927, Gentry submitted an "Application for Pilot's License" to the Aeronautics Branch of the Department of Commerce, passed her examinations, was found physically fit to fly by the department's medical director and soon after was recognized as one of the first nine American women to have obtained a federal pilot's license.

3

A Record Year

1928

I think I inspired the confidence of women in aviation.
—Viola Gentry in James Kilgallen's article "Girl Pawned Gems to Become Flyer and Break Record," Kane Republican, *December 24, 1928*

Although Gentry had applied for, completed and passed the necessary requirements to obtain a United States Federal Pilot License (which included ten hours of solo flight time within two months of submitting the application, successful completion of aerial and ground maneuvers as directed by an inspector and a physical examination), she had yet to receive one. In a letter to the Department of Commerce dated February 6, 1928, which is in the files of the Federal Aviation Administration, Gentry asked, "Will you please advise me when I may expect to receive the license and what is holding it up[?]"

Clarence Young, director of aeronautics, responded to Gentry's letter the day after it was received. He informed Gentry a license had not been issued because her "examination papers could not be located." He further stated a letter would be sent to D.G. Richardson, the instructor who had given her the required examinations, and that a license would be issued once they received his response. Within the week, on February 16, 1928, Richardson replied to Young. He said, "I remember Miss Gentry very well as her flying was A1 for a girl…[I] gladly confirm the fact that she has successfully completed her work for Private Pilot's License. It was…at Curtiss Field last September."

That same day, Gentry was issued a private pilot's license, number 1822. On February 25, the license was mailed, and Gentry received it on the twenty-eighth. It is interesting to note that Gentry received the license in Martinsville, Virginia, at the home of Thelma and Oakley Cabell Hayes, her sister and brother-in-law. The address provided by Gentry on her "Application for Pilot's License" and subsequent correspondence with the Department of Commerce read, "Starling Avenue c/o O.C. Hayes, Martinsville, Virginia."

While residing in Virginia, Gentry was mentioned in the article "Nine Women Pilot Planes with Uncle Sam Approving." The story, which was published in the March 13, 1928 issue of the *Daily Northwestern*, named the nine women who had received a federal pilot's license, along with their places of residence. The pilots were listed in random order, not by the date in which they received their licenses, and brief commentaries were written on three of them. According to the author, the nine licensed female pilots were: Phoebe Omlie, Memphis, Tennessee; Ruth Nichols, Rye, New York; Martha Helen Croninger, Fort Thomas, Kentucky; Ruth Elder [no state listed]; Geraldine Grey, Buffalo, New York; Katherine Schuyler Van Bechten, Colorado Springs, Colorado; Marjorie Stinson, Washington, D.C.; Mrs. Frank H. Fredeman, Little Rock, Arkansas; and Viola Gentry, Martinsville, Virginia.

The precise amount of time Gentry stayed with Thelma and Oakley is not known. However, based on newspaper accounts and correspondence with the Department of Commerce, she lived with them from approximately September 1927 until August 1928.

By September 1928, Gentry had returned to New York, a fact made known in an article about the National Air Race, which began on September 5. On that day, thirty-seven planes departed Roosevelt Field in Mineola, Long Island, New York, on a three-thousand-mile race to Los Angeles. The *Ludington Daily News* ran the announcement "National Air Derby Starts," which stated, "The first plane was started by Miss Viola Gentry of New York who waved a handkerchief from the window of a building near the field when a pistol fired by Mayor George E. Cryer at Los Angeles was heard through a transcontinental telephone conversation."

In October, Gentry was in the news again. Henrietta Gee, whose article "Mistresses of the Sky" appeared in the October 14, 1928 issue of the *Syracuse Herald*, featured photographs of "Three Queens of the Air—Viola Gentry…Louise McPhetridge [Thaden]…and Elizabeth Warner."

When speaking of Gentry, Gee reported, "Flying lessons are so expensive that most working girls believe they cannot afford them. But Viola Gentry,

Left to right: "Princess [Troubetzkoy] secretary" and Viola Gentry, 1928. *Courtesy of the Elmo N. Pickerill Papers, Manuscript Division, Library of Congress, Washington, D.C.*

whose home address is Martinsville, Virginia, determined that lack of money would not handicap her."

It is unclear why Gentry's home address was given as Martinsville, Virginia, although it is possible she was interviewed prior to September. Whatever the reason, the documentation and location of her home address would soon become important—particularly to the city of Martinsville, Virginia.

Between September 1927 and September 1928, there is little record of Gentry's activities. She no doubt remained abreast of women's activities in the air and worked toward her own aviation goals. In 1928, female pilots like Amelia Earhart and Louise McPhetridge Thaden made great achievements in the field of aviation. Earhart, in June 1928, became the first female passenger on a transatlantic flight, while Thaden set a women's altitude record of 20,260 feet on December 7.

Gentry had already made a name for herself as a pilot, but she was resolved to further the cause of women in aviation and to showcase their flying abilities. Henry M. Holden and Captain Lori Griffith, in their book *Ladybirds*, said, in regard to the late 1920s, "Women had been flying for almost two decades and yet the books did not reflect their achievements."

Aware of this fact, Gentry was determined to do something about it; she would attempt to set the first officially recorded women's solo endurance flight record. All who heard of Gentry's plan—pilots, mechanics and field

Amelia M. Earhart, circa 1928. *Courtesy of the Prints and Photographs Division, Library of Congress, Washington, D.C.*

assistants—were more than willing to help. Grace Lyon, a wealthy pilot and promoter of aviation from Long Beach, New York, backed the flight financially; she also loaned Gentry her airplane—a Travel Air biplane with a 125 horsepower Siemens-Halske engine. The article "Woman Hopes to Establish Flight Mark," which was published in the *Geneva Daily Times* on December 20, 1928, reported, "Field mechanics supervised preparation for [Gentry's] flight free of charge."

The article further stated, "Field attendants said Miss Gentry was clad for her flight more warmly than any aviator to ever leave the field. Her equipment included a heavy topcoat, two flying suits, a woolen head protector, a fur-lined helmet, fur-lined gloves reaching to the elbows and fur-lined booties."

Gentry's "equipment" also included a parachute, which, according to *Hangar Flying*, was loaned to her by Lieutenant James H. "Jimmy" Doolittle and was "not exactly comfortable."

On Thursday, December 20, 1928, at 5:44 a.m., Gentry took off from Roosevelt Field. She was resolved to stay aloft longer than any female pilot before her and to have her time officially recorded. The attempt would not have been easy on a good day, but on this day, Gentry had to fight a bitter wind and freezing rain. She had equipped the airplane with one hundred gallons of fuel and intended on flying for thirteen hours, thirteen minutes and thirteen seconds. She, like several other female pilots, believed thirteen to be a lucky number.

After five hours of circling Roosevelt and Curtiss Fields, Gentry dropped a message to a reporter below. An unidentified newspaper clipping, which is in the collection of the International Women's Air & Space Museum,

Viola Gentry is seen just before takeoff on December 20, 1928. *Courtesy of the 99s Museum of Women Pilots.*

provided a transcription of Gentry's note. It said, "I suppose I can speak to more people through the evening press journal than in any other way. So 'Hello' from 2,000 feet over Roosevelt Field. A few more of these cold lonesome hours and will have hung up another record. alt [altitude] 2000 revs [revolutions] 1400/Viola Gentry."

Photograph of Viola Gentry likely taken in Atlantic City, New Jersey, in December 1928. The airplane might be the Travel Air biplane loaned to Gentry by Grace Lyon and the one in which she set the first officially recorded women's solo endurance flight record. *Courtesy of the International Women's Air & Space Museum, Cleveland, Ohio.*

As the hours passed, the weather turned from bad to worse. The rain began to freeze on the airplane's wings, and a thick fog rolled in. Conditions had become dangerous, and Gentry knew she had to land. After a "perfect landing," as reported by another undocumented newspaper clipping in the collection of the International Women's Air & Space Museum, Gentry "leaped nimbly out of her ship, and opened a thermos bottle of coffee, the first food she had touched since taking off at 5:44."

Within two hours of landing, Gentry had written an account of the flight "Exclusively for the *Observer* and the North American Newspaper Alliance." The following day, the article was printed, and on December 22, 1928, her story was featured in the *Charlotte Observer*. The headline read, "Tar Heel Girl Tells Own Story of Record Flight/Viola Gentry, Formerly of Rockingham County, Reveals Hardships and Terrors She Met in Setting Endurance Mark." Beneath the headline, a brief commentary was provided.

> *By flying eight hours, six minutes and thirty-seven seconds through winter weather in an open-cockpit plane, Miss Viola Gentry has broken the official world endurance record for women. Two hours were flown in complete darkness, most of the time in hazy and unfavorable weather and the last part in rain. She had only about 100 hours' experience in piloting an airplane previous to her record-breaking flight.*
>
> *Miss Gentry was timed by Harry Booth…who was deputized for the purpose of by the National Aeronautical association* [sic]. *A sealed barograph, placed in the plane by Mr. Booth, was carried on the flight to record proof of her claim to the record.*

Gentry's account of the flight followed:

> *Roosevelt Field, Long Island, N.Y., Dec. 21.—The first few minutes after I took off…was the worst of it all.*
>
> *Rolling along at 5:44 a.m. I got the ship off easily, even though I had over 600 pounds of fuel and oil in a tank in the front cockpit. I climbed up with nothing but blackness around me. I flew toward Curtiss Field. At 5,000 feet, I switched on the dashboard lights. I got the shock of my life when I saw that my oil temperature was down to zero and my oil pressure gauge showed only one pound pressure.*
>
> *The temperature had been 60 when I took off. This was about right for the Siemens-Halske motor. I spent the next two minutes trying to remember*

what my instructor, Bill Ulbrich, had told me to do in such a case. The clouds had closed in under me completely.

It was so dark all around me that I couldn't tell which was up and which was down. Any pilot who has flown at night without a complete set of instruments can tell you what this feeling is.

The boys on the field told me they were scared at this time, but I can only say that I was more than scared. I finally decided to depend on my intuition. So I said, "Here goes!" and cut the motor down to 1,600 revolutions a minute and started her toward what I thought was down and back toward Roosevelt field.

I watched the altimeter. I put my confidence in it to show whether I was going up or down. The air speed indicator might do, but I have more confidence in the altimeter.

The most wonderful sight I saw in my life was the red beacon on Roosevelt field which appeared to me when I came out of the clouds at 1,000 feet. After that I kept it in sight until daylight.

After the flight, the boys at the field told me they had been nearly scared to death while I was lost. They said that just when they were most concerned about my safety, the morning gun went off over at Mitchel field. Then one of the mechanics jumped in his car and started tearing across the field in the direction of the gun. They thought I had hit the ground and they were looking for blotters and things with which to pick me up.

I don't think I was really excited all this time—that is, not until I saw the beacon. Then I went to work with the wobble pump—to pump my gas from the spare tank into the gravity tank where it could flow to the motor. I was so scared I could hardly work it, but my fright only lasted a minute.

I had a pistol with me, according to the department of commerce rules for night flying. But honestly, if I had had to use it at this time, I don't believe I could have fired it—I was so nervous. But I soon got over it.

I didn't lose sight of the beacon on Roosevelt field any more during the two hours until daylight. I was flying Miss Grace Lyons ship and I didn't want to break it up. I was wearing a parachute. They are all right. I believe in them. But mine was very uncomfortable. And I would have stayed with that ship even if it had started to burn. She had loaned it to me and I was going to bring it down the best I could. I christened it the "Grace Lyons" before the flight.

After I came out of the clouds, nothing was wrong with me, but I was still a little nervous. Then I said to myself: "When the time comes to die, you die. And you don't die but once."

Then I got over my nervousness immediately…

After dawn, I kept flying around, keeping the field in sight, at one time I climbed up to 2,000 feet, but it got terribly bumpy, so I came back to 1,000 feet.

All during the flight, I was careful to do as I had been told. I kept the nose from ever getting low and did not bank too much on right turns. With the heavy load of gas either one of these things might have meant "bad news…"

I had some sandwiches and coffee in the plane. But they were so far back I couldn't reach them from where I sat and still keep good control of the ship. So I didn't eat. Anyhow, I was too interested to be hungry.

The best time of my flying was from daylight until about 10 o'clock. Then a mist appeared and shortly after that it began to rain. I was sorry when the rain dimmed the windshield so I couldn't see anything and I had to set her down.

Fédération Aéronautique Internationale

UNITED STATES OF AMERICA

ANNUAL SPORTING LICENSE ISSUED TO

Miss Viola Gentry

PLACE OF BIRTH Gentry, N.C.

DATE OF BIRTH June 13, 1900

CHAIRMAN

N. A. A. CONTEST COMMITTEE.

1928

LICENSE NO. 450

F. A. I. CERTIFICATE NO. 6530

ISSUED BY N.A.A.

TYPE OF AIRCRAFT Airplane

Viola Gentry

SIGNATURE OF LICENSEE

VALID UNTIL DECEMBER 31, 1928.

Viola Gentry's Fédération Aéronautique Internationale/National Aeronautic Association of the United States annual sporting license, issued in 1928. *Courtesy of the International Women's Air & Space Museum, Cleveland, Ohio.*

I wanted to fly for 13 hours, 13 minutes and 13 seconds because 13 is my lucky number. I was born on June 13.

I landed the ship all right. I have confidence in my ability to land a ship and take if off properly. I don't know much else about flying. It takes you so long to learn. But I know I can do these two things and after that it's mostly your own judgment. Just like when I was lost in the dark this morning, I didn't know what to do according to the rule, so I just used my own judgment…

I had 14 hours instruction at Curtiss field from [1924] *to 1927. I was handicapped because I had to earn my own money working as a cashier in New York restaurants…*

I came to Roosevelt field last year and did office work for the flying school to help pay for my flying time. I also coached my instructor Bill Ulbrich, who is a Dane, in his studies to pass the test for United States citizenship. In exchange for this, he gave me additional flying instruction.

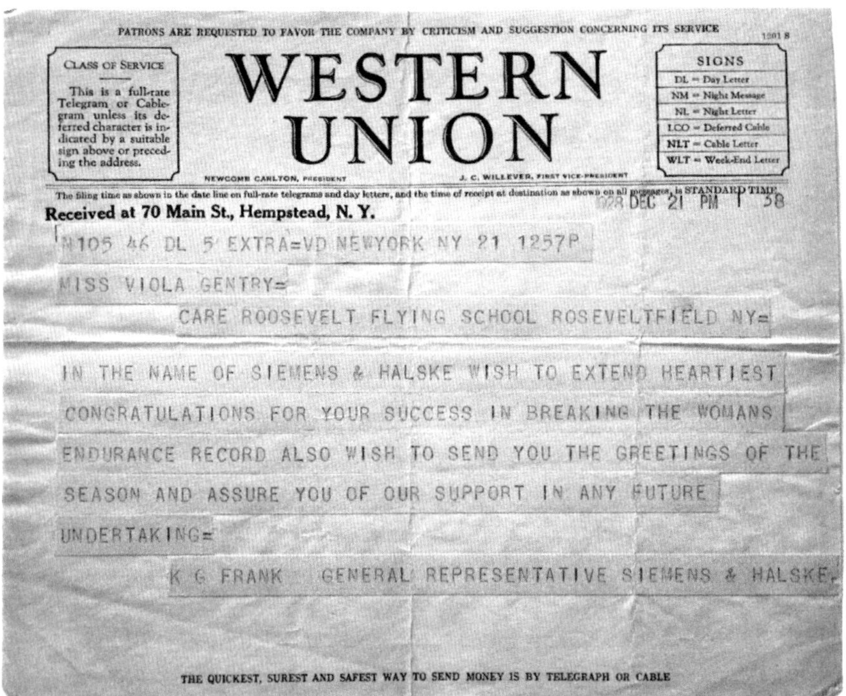

PATRONS ARE REQUESTED TO FAVOR THE COMPANY BY CRITICISM AND SUGGESTION CONCERNING ITS SERVICE

CLASS OF SERVICE

This is a full-rate Telegram or Cablegram unless its deferred character is indicated by a suitable sign above or preceding the address.

WESTERN UNION

NEWCOMB CARLTON, PRESIDENT J. C. WILLEVER, FIRST VICE-PRESIDENT

SIGNS

DL = Day Letter
NM = Night Message
NL = Night Letter
LCO = Deferred Cable
NLT = Cable Letter
WLT = Week-End Letter

The filing time as shown in the date line on full-rate telegrams and day letters, and the time of receipt at destination as shown on all messages, is STANDARD TIME.

Received at 70 Main St., Hempstead, N. Y. 1928 DEC 21 PM 1 58

N105 46 DL 5 EXTRA=VD NEWYORK NY 21 1257P

MISS VIOLA GENTRY=

CARE ROOSEVELT FLYING SCHOOL ROSEVELTFIELD NY=

IN THE NAME OF SIEMENS & HALSKE WISH TO EXTEND HEARTIEST CONGRATULATIONS FOR YOUR SUCCESS IN BREAKING THE WOMANS ENDURANCE RECORD ALSO WISH TO SEND YOU THE GREETINGS OF THE SEASON AND ASSURE YOU OF OUR SUPPORT IN ANY FUTURE UNDERTAKING=

K G FRANK GENERAL REPRESENTATIVE SIEMENS & HALSKE.

THE QUICKEST, SUREST AND SAFEST WAY TO SEND MONEY IS BY TELEGRAPH OR CABLE

A congratulatory telegram was sent from K.G. Frank, general representative of Siemens & Halske, to Viola Gentry on December 21, 1928. *Courtesy of the International Women's Air & Space Museum, Cleveland, Ohio.*

He and Mr. Carl Schneider, a former flyer in the German army, worked all last night to get my plane in shape for the test this morning. They and the other boys who worked so hard to help me on this flight were very sweet.

I quit my cashier job in a New York restaurant the day after Thanksgiving when I learned I would be able to make an attempt at this record.

I am planning to make a try at the world light plane altitude record and the solo endurance record for both men and women. I hope I'll spend the rest of my days flying.

No other narrative of Gentry's record-setting flight, whether by newspaper reporter or historian, gives as much detail and insight into those eight hours as does her own recollection. When reading about her endurance flight, it is important to understand the significance of the undertaking. In 1928, the FAI did not make a distinction between male and female flight records—male pilots in 1928 had established endurance records of sixty-plus hours. Female pilots, particularly Lady Mary Sophie Heath—the first woman to hold a commercial pilot's license in Britain—called on the FAI to create categories specifically for women's flight records. However, there was much debate over the issue.

In *Hangar Flying*, Gentry said flying clubs and organizations called on the FAI to "change, or rather to open, such a classification for women." She further stated, "The fight succeeded at the conference of the FAI held at Copenhagen, Denmark on June 19–22, 1929" and that "The first FAI-recognized International Feminine Endurance Record was established by Maryse Bastie, of France, on July 28, 1929."

Bastie's flight of "26 hours, 47 minutes, 30 seconds was the first official record by the FAI [for women] since no records established before June 22, 1929, could be recognized."

Despite not being recognized by the FAI, Gentry's endurance time of eight hours, six minutes and thirty-seven seconds was a valid record, as the flight had been officially supervised by a representative of the National Aeronautic Association of the United States and timed. In addition to setting the first officially recorded women's solo endurance flight record, Gentry had also bettered the standing endurance flight time set by Lady Mary Sophie Heath by more than thirty minutes. More importantly, as pointed out in Tom Parramore's article "Viola Gentry," "Her ensuing notoriety served as an inspiration for other female aviators who, in the course of the next few months, established a long series of flying records. The celebrity accorded these flights launched the transformation of flying from a male enterprise to one that welcomed both sexes. Viola Gentry was a path-breaking pioneer in women's aviation."

Female pilots viewed Gentry's record as a new challenge—a time that had to be bettered. When speaking of the endurance flight in *Hangar Flying*, Gentry remarked she had "established a new record for women flyers, and gave them a mark to shoot at. And shoot at, they did!"

In truth, Gentry's record did not hold long, but it did yield considerable publicity. Yet not everyone was excited about her newfound fame. The *Bee*, like other newspapers in the country, ran the story of Gentry and her record-setting flight. However, in the *Bee* article "Girl Goes Up to Set Air Record," which was published on December 20, 1928, an addendum was included at the end of the piece. The author stated:

> *Reports received from The Associated Press that Miss Viola Gentry today making a spectacular flight in New York was a daughter of Sam Gentry of this city, resulted in efforts being made to confirm this early this afternoon.*
>
> *A Mrs. Samuel Gentry was reached at a local department store and was asked if the young aviatrix making a name for herself was her daughter. She replied negatively saying that she did not know Miss Viola Gentry, nor a Sam Gentry, but that her name was Maydie Gentry.*
>
> *Enquiries made of other families bearing this name also brought denials of knowledge of a Miss Viola Gentry of Danville.*

The basis for Maydie's denial of Gentry is unknown. Perhaps she was surprised when a reporter appeared at her place of work, the Belk-Leggat Company, and was unsure how to respond to his or her questions. Maybe she did not want to give out information without first speaking to her husband, although her denial of Samuel is a mystery, too. It could have been that Maydie was embarrassed by Gentry's activities, did not see flying as appropriate for women or believed her stepdaughter's actions to be scandalous and capable of bringing shame to the family. No matter Maydie's reason for denying Gentry, it was clear Samuel and Thelma did not share her concerns.

The following day, the *Bee* printed a follow-up article regarding Gentry's connection to Danville, Virginia. The headline stated, "Aviatrix Who Made Record Lived Here," and made known the following information:

> *When enquiries were addressed to a relative yesterday afternoon in efforts to identify the aviatrix as a former resident knowledge of the girl was denied. During the evening an anonymous caller told of Miss Gentry being a former*

> *resident and last night, when the girl's father heard of his daughter's exploit on return from work, identification was readily given and part of her life story told by her father, who expressed pride in her performance. At the same time, the girl flier's sister, Mrs. O.C. Hayes, who lives in Martinsville, was in contact with her sister receiving a telegram, announcing the safe and successful termination of the endurance flight…*
>
> *Mr. Gentry said last night that he had discovered no direct word from his daughter since the flight but said that he knew of her interest in aviation and that he had experienced misgivings knowing the dangers of flying. As a girl, he said, she showed determination and usually accomplished what she set out to do.*
>
> *Mrs. Hayes who also* [admired] *her sister's pluck said that she was expecting a visit from her during Christmas.*

Gentry had indeed "accomplished what she set out to do." Her hard work, patience and strong-willed determination allowed her to fly even when the odds were against it. Shortly after her record-setting flight, on December 24, 1928, James Kilgallen's article "Girl Pawned Gems to Become Flyer and Break Record" appeared in the *Kane Republican*. In it, Kilgallen documented the most substantial burden in Gentry's quest to fly—money. He remarked, "She laughingly told how earlier in her career, she went 'broke' keeping up with aviation so she pawned all of her personal belongings. She even sold her liberty bonds which had been stowed away 'for a rainy day.'"

For Gentry, those sacrifices paled in comparison to the joy she received from flying. In the air, Gentry experienced a freedom that could not be found on solid ground. As she parted the clouds and soared to new heights, Gentry believed there was no limit to her dreams—that each one was attainable. When she gazed down on the fields and across the horizon, she felt connected to her creator. To Gentry, nothing compared to the sensation of flight. It was a feeling like none other; it was an emotion she desired to share with others—particularly young women—so they, too, might be encouraged to take flight. As quoted by Kilgallen, Gentry said, "I am principally interested in the promotion of aviation…I think I inspired the confidence of women in aviation by taking off Thursday with a heavy load, 670 pounds, staying up 8 hours, and coming down without cracking up."

Throughout her life, Gentry championed the cause of women in aviation and welcomed occasions to speak on the matter. Four days after setting the women's solo endurance flight record, Gentry was a featured speaker on a New York radio station. As reported in the *Bee* on December 24, 1928,

"Viola Gentry, one time Danville resident, made her radio debut [this was not her first radio appearance, as she had spoken on WRNY in 1926] last night at 8:30 from WJZ and the blue chain generally. She spoke for less than five minutes and described recent sensations while flying around Roosevelt Field for a new endurance record for women."

Opportunities for Gentry to talk about aviation and her flying experiences were not hard to come by. Reporters, various institutions, ladies' clubs and other organized groups were anxious to hear her awe-inspiring accounts of aerial success. Several cities and two states were not only eager to hear and tell Gentry's story but also to claim her as their own. Upon the successful completion of Gentry's endurance flight, North Carolina newspapers proclaimed her a "Tar Heel Girl"; Danville, Virginia, reported she had "Lived Here"; and Martinsville, Virginia, made known it was "Home of World's Endurance Flyer."

In fact, the city of Martinsville addressed the matter in the December 28, 1928 issue of the *Henry Bulletin*. The article "Martinsville Claims Home of World's Endurance Flyer" declared:

> *Martinsville can proudly establish equal rival claims as being the home of the holder of the new women's world's record for endurance flying…As soon as the news of this record flashed across the horizon and in the newspapers of the country, citizens here little realized that she was…years ago a resident of Martinsville. Danville papers claimed that she was last a resident of that city before entering the air service, but this claim was discredited by Mrs. O.C. Hayes of Forrest Avenue, this city, a sister of the aviatrix who stated to a reporter for this paper that Miss Gentry made her home in Martinsville and not Danville, before going north. Another claim that this is based upon is that last March the Associated Press carried a news story naming the nine women now licensed by Uncle Sam to fly, and at that time Miss Gentry gave Martinsville as her home.*

So proud was Martinsville of Gentry that in the first week of the new year, the city welcomed her with all the pomp and circumstance due their "Daughter of the Air."

4
Triumph and Tragedy

1929

Things happened fast for me during 1929.
—Viola Gentry, Hangar Flying, *1975*

Gentry flew into 1929 with great fanfare and soared in the spotlight of an admiring nation. After setting the first officially recorded women's solo endurance flight record, she received numerous requests to attend ceremonies and other events in her honor. One of the invitations, perhaps the most exciting, came from Martinsville, Virginia.

The president of the Martinsville Rotary Club, Rives S. Brown, sent a congratulatory telegram to Gentry, along with an invitation to attend the club's upcoming banquet. Gentry, who was thrilled by the request, accepted the engagement. On January 2, 1929, the *Bee* announced her upcoming visit. In the article "Viola Gentry Will Fly to Martinsville," readers were told Gentry was expected in Martinsville on the afternoon of Thursday, January 3, and that elaborate plans were being made to "make this a great occasion when Martinsville welcomes her own air heroine back home."

The article further stated that invitations had been sent to city officials and leaders of civic organizations to participate in receiving Martinsville's "Daughter of the Air."

It is interesting to note that on the same day festivities were being planned in Martinsville, Evelyn "Bobbi" Trout, a twenty-two-year-old pilot, was circling her airplane over Los Angeles, California, in an attempt to break Gentry's endurance record. A headline in the January 2 issue of the *Bee* read,

Evelyn "Bobbi" Trout with her Golden Eagle Monoplane, 1929. *Courtesy of the International Women's Air & Space Museum, Cleveland, Ohio.*

"Girl Goes Aloft to Break Viola Gentry's Mark," while the January 3 issue reported, "Trout Girl Smashes Viola Gentry's Mark."

Flying for twelve hours and eleven minutes, Trout had "eclipsed" Gentry's record. Nevertheless, the news did not cloud the atmosphere surrounding Gentry's visit to Martinsville. As a matter of fact, Gentry was delighted to hear of Trout's success. Another article from the January 3 issue of the *Bee*, "Big Reception Awaits Girl at Martinsville," said, "Miss Gentry showed no trace of jealousy in praising the feat of Miss Trout."

The author then quoted Gentry as saying, "That is just wonderful; it will do so much to encourage girls to take up flying."

Ever the promoter of aviation among women, Gentry not only aspired to fulfill her own goals among the clouds but supported other female pilots, too.

On the same page the *Bee* reported Trout's new record, it also ran the article "Viola Gentry May Pass Over Danville on Way to Martinsville Today." There is no doubt the citizens of Martinsville and Danville were proud of the woman who had lived among them and were eager to catch a glimpse of her. The article informed readers that Gentry was expected to fly into Martinsville by way of Richmond and would likely follow the Southern Railway lines, which would guide her over Danville. The reporter could not provide a time for Gentry's flyover, but it can be assumed that Danville residents kept a sharp eye on the sky. It is also probable some of them traveled to Martinsville to greet Gentry in person, as the distance between the cities is a mere thirty miles.

One person, however, one who was most important to Gentry, did not make the trip. The writer of the *Bee* article spoke with Gentry's father regarding her homecoming and reported, "Should [Gentry] pass over Danville, Sam Gentry, her father, will have the first sight of his daughter in the air since she achieved her recent record. He said this morning that he could not get away in order to go to Martinsville where Miss Gentry is to be feted by the community."

It is unclear why Samuel failed to attend the ceremony for his daughter. Although he had expressed pride in Gentry's accomplishment, occurrences in the past might have strained their relationship. It is a fact that will never be known.

Gentry was no doubt saddened by her father's absence in Martinsville, but the exhilarating welcome given her by the city must have overshadowed any sense of letdown. The January 4 issue of the *Reidsville Review* gave a splendid account of the Martinsville celebration in the article "Viola Gentry on a Visit to Sister in Martinsville." It read:

> *Martinsville, Jan. 3—The greatest welcome that has ever been accorded a person in Martinsville was accorded to Miss Viola Gentry, Martinsville aviatrix…Her late arrival from Danville, where she had been forced down by darkness did not in any manner dampen the enthusiasm of one of the largest crowds that has ever gathered in this city.*

> *Immediately upon her arrival she was carried to the Thomas Jefferson hotel where she was the honored guest at a banquet tendered by the Martinsville club. Rives S. Brown, president of the Rotary club presided at the meeting and acted in the capacity of toastmaster. Special music was furnished by a trio of women musicians…before and after the sumptuous banquet.*
>
> *The address of welcome in behalf of the Rotary club, was made by Rives Brown; in behalf of the Kiwanis club, by J. R. Smith; in behalf of the city, by Mayor George A. Brown; in behalf of the women's clubs, by Mrs. Edwin G. Penn and Mrs. J.W.F. Backner. Dr. J.A. Shackleford and Rev. C.M. Wales also made short addresses during the meeting. All of the addresses were in praise of the accomplishments of Miss Gentry.*
>
> *Rives Brown presented Miss Gentry to the gathering and in a pleasing manner the Martinsville aviatrix related her experiences on the day of the record flight. Regret was expressed that Martinsville had made no progress toward the building of a permanent airport…Much emphasis was placed on the value of an airport for all cities in the country, on account of the rapid development of aviation.*
>
> *Miss Gentry received many flowers and other gifts from the organizations of the city, as well as many personal gifts. She was very much impressed with her reception and expressed her appreciation to the gathering. Although the Patrick Henry hotel and the Thomas Jefferson hotel offered their services to the aviatrix, she chose to stay with her sister, Mrs. O.C. Hayes.*

The *Henry Bulletin* featured the front-page article "Rotary Club Entertains Noted Aviatrix" in its January 4 issue, which provided a vivid description of the elaborate welcome given Gentry. It said, "Yesterday the city took on an appearance of a patriotic holiday, the streets being bedecked with American flags, flying from the poles placed in front of each business house, doing honor to Martinsville's own 'Daughter of the Air.'"

On January 8, the same newspaper ran a lengthier article, "Noted Aviatrix Banqueted by the Rotary Club," and reported, "During her stay here a number of Martinsville people took a ride into the clouds with the aviatrix as her guests."

Gentry remained in Martinsville for three days. While there, she must have imagined herself as a character in a fairy tale or in one of her grandfather's stories. A small-town girl who had run away to join the circus, fought boredom as a roller in a cigar factory and worked tirelessly in New York restaurants was now being heralded as a hometown hero. She reveled in her prestige and was inspired to plan additional record-setting flights, flights that would be original and bold. Eager to return to New York, where she could work out the details of her newly conceived ideas, Gentry—along with William "Bill" Ulbrich, who had accompanied her to Virginia—took off from Dillard's Field in Martinsville on January 7 and headed north.

The flight back to New York required several stops, including one in South Boston, Virginia, where Gentry was met by an impressive number of fans. The *Danville Register* article "Aviatrix Starts on Return Flight," which was published on January 8, reported:

> *A large crowd turned out this morning to welcome Miss Viola Gentry, Martinsville aviatrix, to South Boston, when she stopped here for 35 minutes, while en route to Richmond, on her way to Roosevelt field, N.Y.…As she stepped out of her plane she was greeted by Mayor Wilburn, who presented her with a key to the* [town]…*H.D. Elliott of the Kiwanis* [club] *also greeted the flier and* [pre]*sented her with a corsage. Miss Gentry responded and thanked them for their kindness. Most of her words were in praise of the local airport, which she stated was one of the best between Martinsville and New York…Other delegations were present representing the Kiwanis club, Lions club, and also a large delegation of ladies of the town.*

Gentry's dream of making a name for herself in the world of aviation had come true, but setting the women's endurance record had only given her a taste of what could be. On the flight back to New York, Gentry spent a considerable amount of time conceiving new aviation endeavors. First, she decided to apply for a limited commercial pilot's license when she returned to New York, which she qualified for and received that same month. Next, she planned another endurance flight in which she could reclaim the record, and finally, she discussed plans with Ulbrich regarding how she, and they, could further push the limits of flight. The two had endless ideas, albeit no defined plans, which they did not hesitate to share with friends, acquaintances and reporters. In fact, before Gentry touched ground in New York, newspapers reported she and Ulbrich were planning a transatlantic flight to Hamburg, Germany. Whether the two were serious about the idea is not known; even if they were, Gentry's plan to reclaim the women's solo endurance flight record took priority, and she had a sponsor that could make that goal a reality.

On January 8, Gentry and Ulbrich landed at Roosevelt Field in New York. The *Henry Bulletin*, on January 11, informed readers of her safe return. The article "Miss Viola Gentry Reaches New York Safely Tuesday" also reported that Gentry "would make a flight to Wichita, Kansas, where she was to be presented with a $7,000 airplane by the Swallow company."

By January 20, Gentry had arrived in Wichita and accepted the Swallow biplane. While there, Sara Ellen Wilson, an eleven-year-old girl, asked Gentry for her autograph. Gentry, according to a story in an undated clipping of the *Wichita Beacon* that is in a private collection, autographed the young girl's scrapbook with the words "Viola Gentry/I can die only once/Jan 20, 1929." A photograph of Viola Gentry, the autograph and the Swallow biplane accompanied the article.

Because of bad weather, Gentry was unable to fly her new airplane to New York and therefore returned by train. Ulbrich, who had accompanied Gentry to Kansas, remained behind and waited for improved flying conditions. Although Gentry was in Kansas for a short period of time, her determined spirit impressed a number of people. After her departure, Wichitans, as stated in the *Wichita Beacon* article, began following Gentry's aviation endeavors with great interest.

When Ulbrich arrived in New York with the Swallow biplane, Gentry did not waste time preparing it, or herself, for the endurance flight. She made arrangements to load test the airplane on January 31, as it would be carrying an extra sixty gallons of fuel in the cockpit. Because Gentry had never flown an airplane with an additional tank of fuel, she knew it was imperative to get the feel of the airplane before attempting the record.

The day prior to Gentry's test flight, seventeen-year-old Elinor Smith, the "Flying Flapper," took to the sky over Mitchel Field on the Hempstead Plains in Long Island, New York. Smith had set out to establish her own women's solo endurance flight record, and after remaining in the air for thirteen hours, sixteen minutes and forty-five seconds, she did just that. Smith's new record did not discourage Gentry, as she had already planned to stay aloft twenty hours.

On the day of Gentry's test flight, as reported in the article "Viola Gentry in Airplane Crash," which appeared on February 4 in the *Landmark*:

> [Gentry] *roared down* [Roosevelt Field's] *famous transatlantic runway and got into the air just before reaching the treacherous gully in which René Fonck piled up his ship at the start of a flight to France two years ago last September. But she was still low and as she passed over the gully her plane staggered in an upward current of air.*
>
> *At this point she apparently made a mistake in judgment, turning her plane down wind and so losing the lifting assistance of a head wind. This caused the ship to slide and fall.*

Elinor Smith, the "Flying Flapper," April 30, 1931. *Courtesy of the International Women's Air & Space Museum, Cleveland, Ohio.*

Keeping her composure the woman pilot cut her ignition switch, thus keeping her plane from catching fire at the crash…

Miss Gentry's plane was so badly damaged that it will take at least a week to repair it, but she stepped from the cockpit unscathed and apparently unruffled. She blamed the crash entirely on her handling of the plane.

The *Standard Union* ran the article "L.I. Air Girl Unhurt; Rough Wind Blamed" on January 31 and provided further details of the crash. It stated, "The landing gear was crushed, the lower wing badly damaged, and the propeller broken. The motor, however, was undamaged. Miss Gentry was without a scratch."

Newspapers across the country featured photographs of Gentry and her crashed airplane. Headlines and captions read, "Hard Luck for Viola," and "Tough Luck Spoiled the Hopes of Viola Gentry."

Nevertheless, the crash did *not* spoil Gentry's hope of reclaiming the women's solo endurance flight record or dishearten her in any way. In actuality, the crash provided a good laugh for Gentry, as well as membership into an elite club—the Royal Order of Flying Jackasses.

As Gentry said in *Hangar Flying*, "To qualify for membership in this Order, you had to pull a whopper of a boner—not just a tiny one, but a real humdinger! There were no dues and the only person who could say you were eligible was Mr. [L.B.] Rawlings."

Rawlings was a pilot at Curtiss Field who witnessed many foolish mistakes made by pilots. As a result, he created the infamous club. In speaking of the crash and her entry into the group, Gentry further said:

> *Everyone on the field had been telling me what to do and what not to do, and the gist of every bit of advice had been—"Viola, whatever you do, if you have to make a forced landing, for heaven's sake, shut off the gas!"*
>
> *...I shut off the gas all right but ran completely out of experience. I failed to remember that I was not heading into the wind—and forgot to take the drift out of the plane. Sure—I shut off the gas—but in landing, took off the landing gear and smashed the* [propeller].
>
> *Mr. Rawlings immediately announced that I was now a member of the Royal Order of Flying Jackasses and that night presented me with my membership pin—a stickpin set with a small gold jackass.*
>
> *Was my face red! Well, here I had been thinking I was an experienced pilot,* [but] *I sure pulled one whopper of a boner that time!*

While Gentry waited for her airplane to be repaired, she contemplated her next endurance flight and sought aviation-related work. An opportunity for such a position came in February, when E.H. Holmes broke ground on what would become Holmes Airport in Jackson Heights, Long Island, New York. The airport's groundbreaking ceremony and dedication took place on February 14. Among those present were E.H. Holmes, sponsor of the airport project; Clarence Chamberlin, who flew from New York to Germany carrying the first transatlantic passenger; Dorothy Stone, actress and dancer; and Viola Gentry.

Gentry gave a short account of the ceremony in *Hangar Flying*:

> *I was there, in a semi-official capacity, for I was slated to be the "hostess" for the new airport.*
>
> *It was a very cold day and I remember that I was dressed in a long blue leather coat with matching beret. I thought I was just the last word in chic—and then Dorothy Stone arrived. She was wearing a leopard skin coat and a little turban and she was so charming that I just slunk into the background and stayed there.*

She might have wanted to stay in the background, but a photograph of the dedication, published in the February 15 issue of the *Daily Star*, showed Gentry standing in front of the crowd alongside Dorothy Stone. They both held shovels.

Viola Gentry shakes hands with George Haldeman as E.H. Holmes looks on. Photograph possibly taken in March 1929 at Holmes Airport. *Courtesy of the San Diego Air & Space Museum.*

In regard to the position of airport hostess, Gentry would have to wait for the job, as the airport was not scheduled to open until mid-March. In the meantime, Gentry focused on setting and breaking records, but she was not the only female pilot doing so.

On March 3, the *Brooklyn Daily*

Eagle featured an article titled "Endurance Trials by Aviatrices Assume Character of Beauty Contests; Entrants by Score Now Seek Mythical Air Title." O.R. Pilat, author of the piece, began the article in a condescending manner. He stated, "Solo endurance flying for women, which started out as a high, risky adventure, has become a variety of beauty contest with fair young aviatrices vying for the front page of newspapers because of their contemplated or completed flights."

Pilat then addressed the issue of women's endurance records and stated they were "mythical," as the FAI did not recognize them. In speaking of Gentry's and Elinor Smith's records, he claimed that because their records "have been considerably less than those set up by the men, officially the girls have been just amusing themselves." Pilat further explained how pilots like Amelia Earhart were working to "secure recognition for their sex" by establishing their own organization, "just as there now is in tennis and golf and other recognized games."

Near the end of the article, Pilat made a statement about women's flights, which was likely his most profound. He stated, "Like New Year's resolutions, records seem to be made only to be broken, and with as little real benefit. But, the benefit to aviation is that more and more women are taking to the air…Though the endurance record battle at times seemed amusing to witnesses, Miss Gentry's recent unsuccessful attempt reveals that danger is always lurking beneath the surface to give the flight dignity."

It is unknown whether Gentry acquired the hostess position at Holmes Airport, but it is known that she was a frequent visitor there. By mid-March, Gentry's Swallow biplane had been repaired, and on Wednesday, March 20, she announced she would take off from Holmes Airport the following Saturday in an attempt to reclaim the women's solo endurance flight record. Two days later, however, staff at Roosevelt Field reported Gentry would take off from their runway. Whatever the reason for the location change, Gentry should have stuck with her original plan, as she was grounded by Lieutenant H.B. Clarke, manager of Roosevelt Field, on the day of her flight. In the March 23 issue of the *Scranton Republican*, the article "Jensen to Start Air Test Today" explained that Lieutenant Clarke had "forbade [Gentry] to make the attempt unless she had a gasoline dump valve installed in her little plane."

After her grounding in March, Gentry did not fly the Swallow biplane again. The basis for that decision is unknown, but it could have had something to do with a job she acquired in April. Early in 1929, Gentry received a request to deliver a new Arrow Sport biplane to Hugh Morton, an attorney in Boston, Massachusetts. The job was deemed a newsworthy event and was reported on across the country. The *Decatur Evening Herald* published a photograph of Gentry inside the Arrow Sport biplane on April 12, 1929. Above the photograph read the headline "She's Got a Novel Job—She Delivers Planes!" and beneath was the caption "Miss Viola Gentry has something new in the way of jobs for women. She is said to be the first woman ever to deliver an airplane."

On April 8, Gentry, along with her co-pilot, William "Bill" Ulbrich, set out to deliver the airplane. The task should have proved uneventful but instead turned out to be one of her most memorable mishaps. Inclement weather prevented Gentry from leaving Curtiss Field that morning, but at 1:30 p.m., the clouds lifted, and she took off.

After traveling a short distance, Gentry encountered foul weather and decided to return to Curtiss Field. There, she and Ulbrich decided to wait one hour and try again. On the second attempt, she was able to travel farther but ran into thick, low-lying clouds. This meant she would have to fly over Long Island Sound at an altitude of three hundred feet, which was a dangerous proposition. Gentry, not wanting to put herself or Ulbrich at risk, returned again to Curtiss Field. While there, she refueled the airplane and checked the weather report. It was then that an event occurred that set Gentry on the path to her finest misadventure.

When speaking of the incident in *Hangar Flying*, Gentry said, "At the field I was given both the gas and another weather check, together with some snide remarks about it being all you could expect with a girl at the controls. That settled it! Months before I had been told that when the birds stay out of the air, you stay on the ground too, but no one was going to tell me that a girl could not fly any time a man thought he could."

Gentry made up her mind to attempt the flight again. For the third time, she and Ulbrich took off from Curtiss Field and headed north toward Boston. On this flight, however, Gentry decided she would take a different route, and no matter what happened, she would not return to Curtiss Field.

A light rain started to fall when Gentry reached Connecticut. She knew the rain would bring dark skies, so she headed toward an airfield in Worcester, Massachusetts. The airfield Gentry chose had recently installed a beacon, which she knew would be helpful when it came time to land.

Ulbrich plotted the destination on a Rand McNally road map, and Gentry flew low to ensure she was flying in the right direction. By all accounts, Gentry was on the appropriate course and in proximity to the airfield. The sky grew darker, and Gentry began to circle the area. She and Ulbrich scanned their surroundings, as they expected to catch a glimpse of the beacon. They saw nothing.

As luck would have it, Gentry began to run low on fuel and needed a place to land. The airfield, which was nowhere in sight, was no longer an option. Flying low and searching for an area in which to land, Gentry spotted a large open field. There, she put the plane down and, as told in *Hangar Flying*, "landed with a crunch against a small mound (which had a horrible odor)."

Gentry looked around and saw a scattering of houses on one side of the field and large buildings on the other. Not knowing where she was, Gentry decided it would be in her best interest to contact Irwin Keyes McWilliams, manager of the Worcester airfield. She walked to a nearby home in the hope it contained a telephone while Ulbrich stayed behind to guard the Arrow Sport biplane.

The homeowner permitted Gentry use of the telephone. She immediately rang the airfield but was informed that McWilliams was not there. His wife had given birth earlier in the day, and it was suspected he was at the hospital. During her conversation, Gentry noticed that several visitors, including six policemen, had filled the house. She paid them no mind, however, as she imagined they were simply curious about her, the landing and the airplane.

Gentry next phoned the Worcester General Hospital, but McWilliams was no longer there. She left messages at the airfield and hospital with the expectation that he would return her call soon. In the meantime, Gentry called the agency that had hired her to deliver the airplane and informed them of her predicament. The secretary at the agency asked the location of the airplane, but Gentry did not know. She therefore asked a policeman and, as she stated in *Hangar Flying*, "nearly dropped dead when he told me that I had landed on the grounds of the State Insane Asylum—five miles from the airport—and that I would have to be identified by some responsible person before Bill and I could be allowed off the grounds."

Gentry at last understood why the police had come to the house. She, the homeowner and the policemen sat and drank coffee while they waited for McWilliams to call. At 8:00 p.m., McWilliams finally received Gentry's message. He called right away and said he would be out to speak with the policemen as soon as he could contact his lawyer. McWilliams worried there might be legal issues to consider if there was damage to the grounds or airplane.

Postcard of the State Insane Hospital in Worcester, Massachusetts, postmarked 1908. *Courtesy of the author.*

At 11:00 p.m., McWilliams arrived, identified Gentry and Ulbrich and took care of the necessary paperwork. Gentry and Ulbrich, who were quite hungry, were finally allowed to leave. After finding a place to eat and recover from the day's excitement, Ulbrich told Gentry the propeller of the airplane had been damaged and would need to be repaired. It was news she did not want to hear. Ulbrich then informed Gentry the grounds of the hospital were unharmed—all except the pile of manure on which she had stopped the airplane. Gentry roared when she heard the news and found the event to be so funny that she took pleasure in telling the story of her malodorous landing for the rest of her life. In fact, when Walt and Ann Bohrer wrote the book *Tales Up!*, which is a collection of humorous anecdotes from early pilots, they told of Gentry's landing on the manure pile and quoted her as saying she was right where she belonged "on a pile of manure in an insane asylum."

Without a doubt, Gentry's best trait was that she could find humor in any situation and remain positive even when the outcome of an endeavor was not as she had hoped.

As far as one knows, Gentry was neither sued nor held responsible for damages incurred from her abrupt landing.

Two weeks after the incident in Worcester, Gentry began preparing for another endurance flight. The April 21 issue of the *Brooklyn Daily Eagle* announced Gentry would soon attempt another women's solo endurance flight record and that "she would use an Arrow Sport biplane, instead of the repaired Swallow plane in which she 'cracked up' several months ago...She said she expected the plane to be delivered next week."

On April 29, the same newspaper reported, "The Arrow sport biplane... was put through a test flight yesterday at Roosevelt Field. Miss Gentry declared she would make the flight next week and said she hoped to stay aloft 30 hours."

The circumstances under which she received the Arrow Sport biplane for her endurance flight are unknown, but it was likely loaned or purchased by a sponsor. On May 3, the *Brooklyn Daily Eagle* published a photograph with the caption, "Soda pop christening was given to the Arrow sport plane yesterday by little Miss Annette Rockwell, 10. Miss Viola Gentry, aviatrix, will use the plane for a solo endurance flight."

Included in the photograph, but not identified in the caption, was Lee Rockwell, the father of the little girl. Because his daughter christened the airplane, it is easy to assume he planned to sponsor this particular endurance flight. Yet despite the photograph and newspaper announcements, there is no documentation Gentry ever attempted an endurance flight in the Arrow Sport biplane.

In actuality, Gentry and Rockwell likely saw an opportunity to go after a bigger record, a record that would bring greater acclaim, and decided to pursue it instead. After the incident in Worcester, Gentry became focused on setting a refueling endurance flight record. Unlike her first record, this type of flight required her to stay in the air, day after day, until the latest record had been beaten. She could not land for fuel but, instead, had to retrieve it in flight. To do this, some pilots flew low and used a device to pick up tanks of fuel from the ground, while others refueled through the assistance of a second airplane. Using that technique, a refueling airplane would position itself above the airplane needing fuel and drop down a hose. Fuel would then be pumped from one airplane to the other. A pilot could not fly and retrieve gas at the same time; therefore, it was necessary to have two pilots onboard.

The same week Gentry posed for photographs with the Arrow Sport biplane, she came into possession of a Paramount Cabinaire biplane with a Warner 110-horsepower engine. Walter J. Carr—a pilot, airplane designer and creator of the Cabinaire—offered the airplane to Gentry, as well as his services as co-pilot, for a refueling endurance flight attempt. According to Robert F. Pauley, in his article "The Paramount Aircraft Corporation," the Cabinaire had a closed-cockpit cabin, which Carr realized early in his career was essential because "if aviation was ever going to become practical, the pilot and passengers had to be seated in a warm, comfortable cabin and not exposed to the elements."

The Cabinaire was ideal for a refueling endurance flight, and Gentry was ecstatic to have an opportunity to attempt something so daring.

At the time Gentry was planning her flight, the refueling endurance flight record was held by the United States Army Air Corps. From January 1 through January 7, 1929, a crew of five men flew an Atlantic-Fokker C-2A near Los Angeles, California, for 150 hours, 40 minutes and 15 seconds. This was the time Gentry would have to beat.

The name given to the Atlantic-Fokker C-2A airplane was *Question Mark*. When asked how long they could stay aloft, the crew was unable to give an answer, hence the name *Question Mark*. Gentry, in response to their flight, proudly named her Cabinaire *The Answer*, as she was determined to stay in the air longer.

Gentry's intention to establish a new refueling endurance flight record received national attention. On May 20, 1929, the *Bee* published the article

The *Question Mark*, an Atlantic-Fokker C-2A airplane, refuels over Southern California, 1929. *Courtesy of the Prints and Photographs Division, Library of Congress, Washington, D.C.*

"Viola Gentry Is Ready for Big Flight," which declared, "Martinsville Woman Plans Attempt to Smash 'Question Mark's' Record."

The article further stated:

> *Miss Gentry's announced intention of attempting this endurance flight came simultaneously with a similar announcement on the part of a crew headed by Lieutenant H.B. Clarke, former manager of Roosevelt Field.*
>
> *Lieutenant Clarke said he planned to take off in a monoplane today, in an effort to stay aloft 200 hours or more. He said Martin Jensen, holder of the world's solo endurance flight record, and a young German pilot, William Ulbrich, who accompanied Viola Gentry in her flight to Martinsville…would accompany him. The plane is called the "Three Musketeers."*
>
> *Their plane will be fueled by means of a device which will enable them to pick up cans of gasoline from the ground.*
>
> *According to Miss Gentry's plans, she plans to hop off Tuesday morning, providing no unforeseen delays occur.*
>
> *…Refueling of "The Answer" will be from another plane…A special pumping device has been installed in "The Answer" to permit lifting gasoline from the refueling plane.*

Ever supportive of their "Daughter of the Air," the article ended with the following regards, "Miss Gentry's Martinsville friends are anxious to see her win out in her eager attempt to establish another world's record."

Clarke, Jensen and Ulbrich took off from Roosevelt Field in a Bellanca airplane on Tuesday, May 21. After six hours in flight, the crew was forced to land due to damage on the tail of the airplane. Gentry and Carr, who had also planned to takeoff on May 21, did not.

On May 22, the team of Clarke, Jensen and Ulbrich took off once again. Newspapers kept up with their progress, and on May 23, the *Yonkers Statesman* gave a report in which Gentry was the main topic. The article "Aviatrix Cooks for 3 in Flight" said:

> *Roosevelt Field, L.I., May 23. Viola Gentry, Long Island aviatrix, appeared at the flying field here today in a new role—that of cook for the three flyers who are attempting to set a new endurance record in their plane, the "Three Musketeers."*
>
> *The birdmen, looking through the windows of their ship watched Miss Gentry as she made coffee, boiled eggs and toasted bread.*

The preparation of Miss Gentry's first meal was interrupted by a note dropped from the plane in which the flyers demanded "a little service."

Whether in the air or on the ground, Gentry was involved in any and all events that occurred at Roosevelt and Curtiss Fields. Although her own ambitions were often thwarted, Gentry never lost the desire to help other pilots succeed in their endeavors, even if it meant preparing breakfast. Because of her encouraging and helpful nature, Gentry became friends with many of the world's greatest and best-known names in aviation.

By June 7, whatever problem Gentry and Carr had faced with their flight plans had been remedied. On that day, she and Carr took off from Roosevelt Field to test the Cabinaire. When they landed, it was reported the airplane had performed well, but a date for the refueling endurance flight was not given. The Cabinaire was ready to go, but Walter Carr was not. Diagnosed with pneumonia, he was unable to fly.

Gentry did not want to delay the flight and began to search for another pilot. Joseph R. James, who traveled the country performing aerial stunts, was available and agreed to fly with her. On June 19, the *Bee* ran the story "Two Groups Prepare for Long Flights," in which it was announced that "Miss Viola Gentry and Joseph James, using a Cabinaire (CQ) [closed quarters] biplane, will seek to set a re-fueling endurance record for light planes."

It was not to be, however. James had been searching for a stable job, and it just so happened, as Gentry said in *Hangar Flying*, "he got one." Once again, Gentry was tasked with finding another pilot. She would also have to endure a longer flight. From May 19 through May 26, Reginald "Reg" Robbins and James Kelly flew their airplane *Fort Worth*, a Ryan B-1 Brougham, around Meacham Field in Fort Worth, Texas, for 172 hours, 30 minutes and 1 second. This bettered the record set by the *Question Mark* by 21 hours, 51 minutes and 10 seconds.

Never discouraged, Gentry enlisted the assistance of another pilot, and they soon took off to beat the record. On June 21, the *Kingsport Times* announced their departure. The article "Girl and Man Bid at Flying Record" reported:

Roosevelt Field, N.Y., June 21 (AP).—Viola Gentry, former holder of the woman's solo endurance flight record, and Charles W. Parkhurst, former instructor at Roosevelt Field flying school, took off at 11:25:15 [a.m.] *o'clock Eastern daylight time, today on a refueling endurance flight.*

The flight began well, and all went as planned until 9:00 p.m., when Gentry was forced to land. The *Schenectady Gazette* article "Woman Aviator Fails to Break Endurance Mark," which was published on June 22, reported what went wrong:

> *Roosevelt Field, N.Y., June 21 (AP).—Viola Gentry and Charles W. Parkhurst, landed their plane "The Answer" tonight after an unsuccessful effort to break the endurance refueling record. They came down at 9 p.m. eastern daylight time, having been in the air nine hours, 37 minutes, 45 seconds.*
>
> *Hazy weather, which made contact with the refueling plane impossible, was held responsible for the failure. Emil Bergin, pilot of the refueling plane, said the weather was so murky that it was difficult for the two pilots to see each other at a safe distance apart.*
>
> *He also said that way was so "bumpy" the maneuver would have been dangerous in clear weather.*
>
> *Miss Gentry, who is known as the "flying cashier"...said after alighting that she will practice refueling during the next few days and make another attempt.*

Other newspapers ran disparaging headlines and comments about Gentry's refueling attempt. The *Dunkirk Evening Observer* published an account of Gentry's flight in its June 22 issue under the headline "Viola Gentry's Effort at Endurance Flight Record Complete Failure," while the *Davenport Democrat and Leader* provided the following commentary on June 23:

> *Viola Gentry, ambitious woman flyer, and Charles W. Parkhurst, took the air over Roosevelt Field Friday for the purpose of attacking the refueling record...The Gentry plane had been named The Answer. It landed after about nine hours. A short answer, Viola.*

For the next few days, Gentry practiced and prepared for another flight. She was now more than certain she could break the record and was eager to get back in the air. At the same time, Parkhurst received and accepted a job offer to fly for a corporation, a position he had long wanted. Once again, Gentry was in need of a pilot.

At the recommendation of Herbert McCory, a newspaper reporter, Gentry contacted Clyde Pangborn in New Jersey. Pangborn was a stunt

Left to right: Clyde Pangborn and Duke "Diavalo" Krantz, August 1, 1925. Pangborn, a stunt pilot and co-founder of Gates Flying Circus, recommended Gentry ask John W. "Big Jack" Ashcraft to co-pilot *The Answer*, as "there was no better pilot in the game." *Courtesy of the National Photo Company Collection, Prints and Photographs Division, Library of Congress, Washington, D.C.*

pilot and co-founder of Gates Flying Circus. According to Gentry in *Hangar Flying*, Pangborn "suggested Jack Ashcraft, who had been barnstorming with Gates Flying Circus," as "there was no better pilot in the game."

John W. "Big Jack" Ashcraft was born in Oklahoma in 1896. At a young age, he and his family moved to Protection, Kansas. He served in World War I and afterward, because of his interest in aviation, moved to the southern United States, where work was available in that field. Later, he performed death-defying aerial stunts for Gates Flying Circus, worked as a flight instructor for Gates Flying Service and opened aviation businesses in Towanda, Pennsylvania, and Macon, Georgia. Ashcraft was considered an expert, daring and safe pilot.

Left: John W. "Big Jack" Ashcraft, early to mid-1920s. *Courtesy of Allen Keller and Cindy Weigand.*

Below: Postcard of Gates Flying Service airplanes and hangars at Holmes Airport in Jackson Heights, Long Island, New York, circa 1929. *Courtesy of the CardCow Vintage Postcards and Collectibles/ Public Domain.*

Ashcraft, like Gentry, had seen and experienced a number of flying-related mishaps, some of which had tragic consequences. In 1924, while flying in Baton Rouge, Louisiana, Ashcraft's airplane went into a tailspin and crashed. The passenger was killed, and Ashcraft was badly injured. Four years later, on January 24, 1928, while piloting an airplane in Macon, Georgia, Ashcraft's passenger climbed out of the cockpit, stood on the wing and jumped. In the article "Pilot Says Man Jumped from His Plane to Death," which was published in the January 25, 1928 issue of the *Chillicothe Constitution-Tribune*, Ashcraft quoted his passenger—Grantland Irwin—as saying, "It's all over now."

The following month, on February 18, 1928, Ashcraft was witness to what was likely the most horrific event in his life. Ashcraft's younger brother Francis and good friend Samuel "Buck" Steele were performing at the Southeastern Air Derby in Macon, Georgia. The two men flew high above the crowd, tossing out aerial bombs that exploded loudly in the sky. As explained in the *El Paso Herald* article "Two Killed When Plane Falls in City Street," on February 18, 1928, "The aviators were tossing out the bombs…from an altitude of about 7,000 feet. One of the missiles appeared to have struck a wing of the plane and exploded prematurely, instantly killing Steele. Minus a guiding hand at the controls the craft went into a tail spin…while hundreds…looked on in horror."

The *Towanda Daily Review* gave further details of the event in the article "Flyers Killed in Macon/Buck Steele and His Mechanic/Victims of Accident with Bomb," which was published on February 20, 1928. It stated:

> *A third man was fatally injured when the wreckage of their plane struck him as it plunged 7,000 feet through the air to the main street of Macon, Georgia. Three other persons were seriously injured and a dozen slightly hurt when a sidewalk caved in under the weight of the crowd that rushed to the scene of the crash…Jack Ashcraft, pilot of Towanda's plane "The Spirit of Ammonia" in the national air race from Roosevelt Field, N.Y., to Spokane, Washington, last summer, was in Macon taking part in the aviation meet and saw his brother and Buck Steele, his pal, meet their tragic death.*

The event no doubt enveloped Ashcraft in a cloud of sorrow. But he was a flier, and there was no keeping him out of the sky.

Ashcraft not only had a reputation for being a skilled pilot, but he was also known for his courage. The *Wellsboro Gazette* of Pennsylvania documented Ashcraft's brave nature in the article "Popular Pilot Badly Burned," which was published on August 1, 1928. The article stated that two pilots near a flying field in Warwick, Rhode Island:

> *Were instantly killed…when their plane went into a tail spin at an altitude of about 100 feet, and burst into flames…As the plane fell, the 100 visitors at the flying field, employees and aviators rushed to the ruins in an attempt to save the two passengers. The heat of the fire drove them back, but Jack Ashcraft…member of the Gates Flying Circus, dashed into the inferno in a mad attempt to open the cabin door. He was beaten back by the fire, his arms severely burned.*

In the spring of 1929, Gates Flying Circus gave its final performance. Increasing federal regulations regarding aviation, along with continued maintenance of older aircraft, decreased the profitability of flying circuses. In June, Ashcraft, along with two other Gates performers—William C. "Whispering Bill" Brooks and Ive McKinney—decided to base themselves at the newly constructed Holmes Airport. There, they would offer their services as commercial pilots. The *Brooklyn Daily Eagle*, in the article "Three Musketeer Airmen Who Barnstormed Country in Flying Circus Now Settled as Commercial Pilots at Jackson Heights," which was published on June 9, 1929, announced the men's intentions. The article stated, "Veterans of a thousand stubble fields from which they flew together during the rollicking, romantic days of aerial barnstorming, these modern prototypes of Porthis, Athos, and Aramis have set down their winged steeds and await some new adventure."

Ashcraft did not have to wait long for a "new adventure," as Gentry took Pangborn's advice and contacted him regarding the refueling endurance flight. Ashcraft agreed to meet Gentry at Roosevelt Field to discuss the idea. Once there, Gentry showed him contracts they would receive if they could break the record by two hours. If successful, the flight would bring Gentry and Ashcraft considerable fame and perhaps a bit of fortune. Ashcraft piloted the Cabinaire to see how it handled and flew with Gentry to assess her flying skills. After careful consideration, Ashcraft accepted the offer, as he was certain he and Gentry could set a new record.

In regard to the flight, Ashcraft asked that the refueling plane be flown by friends of his in Hartford, Connecticut. Because Ashcraft had been flying longer than Gentry, and because of his reputation, she did not question his request. Gentry said in *Hangar Flying*, "I told Jack that he was in command and that I would follow any instructions that he might give me."

While making plans for their flight, word circulated that a man and woman, who were not married to each other, would be flying together for an extended period of time. Some found the situation inappropriate and voiced their opinions. Gentry spoke of the matter in *Hangar Flying*. She said, "The newspapers had been publishing stuff about the projected flight and had been deluged with 'crank' letters. I read letters from people who wanted to know how in the world a man and a woman could stay up in an airplane all by themselves, day after day, with no privacy, etc."

Gentry revealed that she and Ashcraft would remedy the situation by placing a plywood shelf against the wall of the cabin. The barrier would be put in place whenever she or Ashcraft slept or used the toilet, thus providing an ample amount of privacy. The flight was well planned, and the two pilots were ready to go.

On June 27, 1929, Gentry and Ashcraft took off from Roosevelt Field with the intention of flying 174 hours or longer. The two had friendly competition, however, as the *3 Musketeers* was back in the air. The airplane had taken off from Roosevelt Field on June 26 with three crew members: Martin Jensen; his wife, Marguerite "Peg" Jensen; and Bill Ulbrich. Gentry did not begrudge the group of fliers, as they were her friends, although she might have enjoyed seeing them prepare her breakfast this time.

The Answer took flight at 8:49 p.m. It did not carry a heavy load of fuel, only 116 gallons, as Gentry and Ashcraft planned to refuel twice a day. Food was also kept to a minimum; each pilot had a ham sandwich, a lettuce and tomato sandwich, a few oranges, coffee and water. Like fuel, food would be delivered daily.

An account of the flight, given by Gentry in *Hangar Flying*, best relates the day's turn of events:

Undocumented newspaper clipping of Viola Gentry and John W. "Big Jack" Ashcraft, June 1929. *Courtesy of the International Women's Air & Space Museum, Cleveland, Ohio.*

When we took off, the weather was beautiful…As we were getting into position for take-off; Jack said "Gee, I forgot to send my mother a telegram. Throttle back and wait here while I run send her a wire to let hew [sic] *know we are taking off."*

He came back…we went up…and I flew my two hours on and then Jack flew his two hours. About two o'clock [a.m.] *the ground fog started to come in off the ocean. Of course I knew what that meant on Long Island because I had been flying there since 1924. Jack had been in and out of Curtiss Field often, but he wasn't as familiar with the territory as I was, so I asked, "Don't you think we had better land?"…*

Jack replied, "Oh, my partner will get gas to us and I'm not worried, but if you are worried, we'd better land and you can get another pilot, because there's no sense of flying with you if you are worried."

I assured him that I wasn't going to worry…

Jack, like myself, wanted to establish a good record. He had secured the refueling pilot, and the ship that would do the refueling was a Fokker…Jack was positive…his friend would get over to gas us up.

By five o'clock [a.m.], *the ground was out of sight and the fog was thick…We got a last look at the ground shortly before five and Jack took over the controls—it was his turn to fly. I got in the back and was ready for a nap when he called me and said, "Now, I want to teach you something. In a case like this, just before everything disappears, you make yourself a square to fly in. Fly for five minutes in this direction—make a left on the compass, and then fly five minutes on that course—and so on, keeping a perfect square going so that you are ready at all times and are able to land in one of these potato patches around here."*

I hadn't thought of it, and it was good of him to tell me…The fog was letting up a bit and I asked him again if he'd like to land—we could always start over—tomorrow would be another day.

But Jack said "Nope. I've got things to do, after we establish this record, but I'd like to establish a really good record." He then went on to say that there was no use worrying about the refueler, he knew the ship was coming down all right to refuel us.

By 5:30 [a.m.] *there was no sight of the refueling plane and we had very little gas. I had been told twice and I did not intend to hear Jack tell me again that the refueler would get to us and ask if I was afraid. At six o' clock* [a.m.] *we were completely out of gas and still at fifteen hundred feet. Jack told me to fasten my seat belt and hold on.*

The Answer headed toward the ground. Ashcraft tried to control the airplane, to set it down in an open field, but the sea-like fog covered the land and hid it from view. A water tank loomed below, its whereabouts unknown. Ashcraft came close yet somehow dodged it as the airplane continued its descent. Beyond the water tank was Hicks Nursery, a large area dotted with small shrubs and plants, a parcel that might prove forgiving to a forlorn airplane. In an instant, the ground and airplane connected. *The Answer* had crashed, according to Gentry in *Hangar Flying*, into "the only tree [a hickory]...around that territory...big enough to damage the plane."

The chronometer stopped at 6:13 a.m. Ashcraft was dead, and Gentry, who lay inside the craft like a cast-down rag doll, glided toward death's

"Death is 'Answer' to Plans of Man and Woman Pilots Seeking Endurance Record," read a headline in the June 30, 1929 edition of the *Roanoke Times*. Photograph shows the crash of *The Answer* at Hicks Nursery in Old Westbury, Long Island, New York. John W. "Big Jack" Ashcraft was killed, and Viola Gentry was severely injured. *Courtesy of the Queens Borough Public Library, Archives, Joseph R. Burt Sr. Photographs.*

door. Ashcraft's friends, the men he had requested to crew the refueling plane—Carl Dixon and Claude Kincaid—did not immediately hear of the crash. Because of the fog, they had been unable to leave the airfield in Hartford, Connecticut, until 8:45 a.m. When they landed at Roosevelt Field, they were told of Ashcraft's death.

Afternoon headlines conveyed news of the crash to the interested public while photographs, accompanied by the words of reporters and eyewitnesses, provided its gory details.

As reported in the article "Viola Gentry Hurt, Man Dead in Plane Crash," which was published in the June 28, 1929 issue of the *Miami Daily News*, "The plane landed on its nose, the motor buried in the ground. Ashcraft's body was wedged between the motor and the ground. Miss Gentry was found on top of the gas tank with one foot tangled up in the wreckage. She was bleeding profusely and was hysterical, lapsing into unconscious on her way to the hospital."

The *Evening World*, in its June 28 article "Jack Ashcraft Killed, Viola Gentry Hurt in Endurance Test Crash," said, "Hicks [Henry Hicks owned the nursery] and his son, Edwin, found the plane splintered against the tree, not far from his greenhouses. Ashcraft was at the controls. Miss Gentry was in the rear…of the plane, separated from the pilot by a curtain [plywood partition] which she had put up before the flight. Ashcraft was dead. The young woman was bleeding from the mouth, but managed to blurt out her name as Hicks bent over to listen."

The *Bee* featured the headline "Viola Gentry Injured in Plane Crash" on the front page of its June 28 issue. The article delivered the dire news that "Miss Viola Gentry, endurance flyer, was probably fatally injured early today when her plane, 'The Answer,' cracked up near the Jericho Turnpike, near Westbury, Long Island."

Prior to the article's publication, the *Bee* had contacted Gentry's father in Danville and her sister in Martinsville to ensure they knew of their loved one's misfortune. Neither had heard of the tragedy. The paper also procured statements from Samuel and Thelma for an additional story, which was printed beneath the abovementioned article. Titled "Viola Gentry's Father Says He Tried to Get Daughter to Give Up Aerial Activities," the piece told of Samuel's "considerable concern" for his daughter and how he had tried to convince her to give up flying after she set the women's solo endurance flight record. The article stated, "He tried to dissuade her from continuing her aerial activities pointing out to her the dangers of flying and the possibility of [a] fatal accident."

No matter his words, Samuel said, "It did not seem much good to try and get her to stop flying."

Thelma, it was said, "registered considerable anxiety on being told of the crash." She had spoken to Gentry the evening before and said her sister "seemed so full of confidence of setting a new record and talked…in a much enthusiastic way."

In regard to traveling to New York to visit Gentry at the hospital, Thelma declared "that she would immediately get in touch with the…hospital and if necessary would go."

Samuel said, "It would be impossible for him to go."

In addition to her family, Gentry's friends were equally concerned. The crew of the *3 Musketeers* heard through radio conversations that the *The Answer* had gone down. The Jensens were aware of Ashcraft's and Gentry's fate but kept the news from Ulbrich. Prior to the crash, there were only a few people who were aware of Gentry's and Ulbrich's relationship. Afterward, their love for each other was revealed in black-and-white splendor. One headline in the June 28 issue of the *Miami Daily News* read, "Crash Bares Love of Aviatrix for Bill Ulbrich, Also Flier." The article reported how Gentry had called out for Bill as she was being removed from the wreckage and that he "was as hysterical as Miss Gentry had been and…pleaded and stormed and wept for a true report on what had happened."

The article further stated:

> *Pilots at the field talked with Ulbrich and finally told him…that Miss Gentry had broken an arm. He was so distracted they were afraid to tell of multiple other breaks, of cruel lacerations and of internal injuries that had made hospital physicians shake their heads and murmur of "a 50-50 chance."*
>
> *Ulbrich was promised, to quiet him, that a microphone would be run directly to Miss Gentry's bedside so he might speak with her. Preparations were made to do this on the chance that in a lucid interval the woman flier might be able to say a word, perhaps her last word, to the man she continued to call for in her semi-conscious moments."*
>
> *Ulbrich lives near the flying fields with his mother and for several months Miss Gentry has been staying with them. She and Ulbrich were often seen together, at the fields and elsewhere, and they were always "Bill" and "Viola" to each other. But that was easily accounted for by the natural intimacy of persons in the same hazardous profession.*

After seventy hours and thirty-eight minutes, the *3 Musketeers* was forced to land, as the crew was unable to refuel. They, like Gentry, had failed at setting a new refueling endurance flight record, but their landing on June 29 was likely for the best, as it is doubtful Ulbrich could have remained in the air much longer considering his state of mind. As soon as the *3 Musketeers* landed, Ulbrich departed for the hospital. When he arrived, however, he was not allowed to see Gentry. Her condition was fragile, and she was not permitted visitors. Ulbrich—along with Gentry's family, friends and fans—could only wait to see if she would live or die.

In regard to Gentry's injuries, newspapers across the country gave varying accounts, but the consensus was that her "condition was critical." After being pulled from the wreckage, Gentry was taken to the Nassau County Hospital in Mineola, Long Island, New York. The hospital was unable to give an immediate report of her injuries but did send out a request for blood.

The story "Airman Killed, Girl Is Injured," which was published in the *Post-Crescent* on June 28, stated:

> *All Offer Blood*
>
> *In the face of tragedy, the spirit of aviation revealed itself today as a spirit* [of] *which all humanity might well be proud.*

Postcard of the Nassau County Hospital in Mineola, Long Island, New York, postmarked 1932. *Courtesy of the author.*

> *A message was sent from the hospital to the flying field that the woman pilot needed blood. Who would volunteer?*
>
> *And as one man, the flying and "ground" staffs stepped forward. Pilots whose names are often in the papers, office executives and their clerks, grimmy* [sic] *"grease balls" who oil and tend the engines, not a man on the field but begged for the chance to help Miss Gentry.*
>
> *Doctors said that as soon as preliminary blood tests were made a donor would be selected and the transfusion operation performed.*
>
> *Word that a transfusion was necessary was also received by members of the Gates Flying circus, to which Ashcraft was attached. Edward* [illegible], *a pilot, and W.M. Weeks, a mechanic, tumbled into a plane at...Holmes airport and flew...a distance of 20 miles and then took an automobile for the hospital in an effort to be among the first to offer assistance.*

Another man who heeded the call for blood was Wallace "Wally" Bishop, a cartoonist and creator of the syndicated comic strip *Muggs and Skeeter*. An amateur flier, Bishop frequented Roosevelt Field during the summer of 1929. According to Gentry in *Hangar Flying*, "Wally Bishop was one of the first to respond when told I needed blood transfusions. They had tested several persons and Wally's blood was the only one which matched mine."

The following day, June 29, newspapers continued to report on Gentry's condition. A headline in the *Prescott Evening Courier* gave the alarming news "Gentry Girl Is Nearing Death." The article stated, "After a blood transfusion late yesterday, Miss Gentry failed to rally and at midnight was reported sinking. She recovered for several hours, however, but shortly before noon her physician said she had only one chance in three to live."

Gentry's injuries—which were eventually confirmed in newspaper articles as consisting of blood loss, shock, a crushed and fractured right arm, a deep gash to her face and internal wounds—presented a less than favorable outlook. Then, on July 1, good news was reported. Her condition had improved, and recovery was expected. Gentry knew the road to a full recovery would be long and full of hurdles. Yet she was ready to begin the journey, as she knew she would have to reach road's end in order to fly again.

In the days following the crash, newspapers reported Gentry would likely not fly again and printed stories regarding a "jinx" that had followed her aviation attempts. On July 2, the *Daily Star* ran the story "Jinx That Dogged Flights of Girl Flyer in Her Pursuit of Record Finally Won" and, in flowing melodrama, told of Gentry's earlier mishaps. The article said she was known as the "tough luck kid of the flying fields" who had twice been defeated by "an enemy, a murderous, invisible enemy—a dreadful jinx of the air."

At the end of the article, the author announced, "The…jinx…won." In order to win, however, an opponent must lose or forfeit his or her game, and Gentry was neither ready nor willing to concede to her "jinx."

By July 6, the wreckage of *The Answer* had been removed from Hicks Nursery; John W. "Big Jack" Ashcraft had been buried in Protection, Kansas; and Gentry had settled into her room at the Nassau County Hospital.

In August, Gentry's condition was deemed stable, and she was at last allowed visitors. Her "Auntie" Sears came down from Bridgeport, Connecticut, while her sister, Thelma, drove up from Martinsville, Virginia. Bill Ulbrich and his mother were frequent visitors, as were a host of fliers—both men and women—from Roosevelt Field. Two fliers in particular—Lewis A. Yancey and Roger Q. Williams, who had recently flown from Maine to Rome—were told they could visit Gentry anytime they wanted, even outside of visiting hours.

Along with an array of visitors, Gentry also received a host of get-well cards and letters. One note, perhaps the most unexpected, was from George A. Gray, the pilot with whom she had first flown. George, along

Wreckage of *The Answer*, July 1929. *Courtesy of the National Air and Space Museum (NASM 88-16086), Smithsonian Institution.*

Wreckage of *The Answer*, July 1929. *Courtesy of the National Air and Space Museum (NASM 88-16086), Smithsonian Institution.*

Wreckage of *The Answer*, July 1929. *Courtesy of the National Air and Space Museum (NASM 88-16086), Smithsonian Institution.*

Wreckage of *The Answer*, July 1929. *Courtesy of the National Air and Space Museum (NASM 88-16086), Smithsonian Institution.*

with his wife, "Jack" Stearns Gray, penned a letter to Gentry, which is in the collection of the International Women's Air & Space Museum, that said:

> *My dear Miss Gentry—*
>
> *We were dreadfully sorry to hear of your "bad" break and do hope you are recovering rapidly.*
>
> *We saw an account in a newspaper sometime ago where you said "George Gray" whom you first flew with at the Ostrich Farm in Florida was killed in the World War. Well, Miss Gentry I am very much alive, and am still in the game, though not the flying end…How in the world did you hear I was dead?*
>
> *We shall keep ourselves posted as to your condition, and when you get well, drop down this way sometime and see if I am not the same old pilot who took you aloft…*
>
> *With all good wishes for an early recovery—and luck—and more luck—*
>
> *We are,*
> *Most Sincerely,*
> *George A. Gray*
> *"Jack" Stearns Gray*
> *Washington, D.C.*

Gentry must have been surprised and pleased to receive a letter from the man who had first carried her to the clouds.

Although Gentry was on the road to a physical recovery, newspaper reporters begged the question of whether she would recover mentally. Articles explained how the "psychological effect" of an airplane crash, in addition to a near-death experience, might bring a pilot's flying days to an end, particularly if he or she did not immediately return to the air. However, writers of such articles could not have known Gentry or her determined spirit.

On September 3, a photograph of Gentry lying in a hospital bed with her head bandaged and a wide smile on her face appeared in the *Daily News*. The caption for the photograph read, "Denies She's Through as Flier," along with a quote from Gentry that said, "I'll fly again…I don't want my friends in aviation circles to think I'm a quitter."

The next month, on October 11, Gentry left the Nassau County Hospital. It was the first time since the crash. She had not been released from the hospital but had been permitted a short car ride with friends. Out of all the places she might have gone, Gentry's only request was to visit Roosevelt Field.

By the end of October, Gentry was released from the Nassau County Hospital. However, as the October 23 issue of the *Standard Union* put it, Gentry was "Out of Danger, into Debt."

Gentry's four-month stay at the Nassau County Hospital had amounted to a small fortune in fees. Grace Lyon, who had sponsored Gentry's first endurance flight, gave $3,000 toward the bill, but there was still more to be paid. The hospital contacted Lee Rockwell, sponsor of Gentry's refueling endurance flight, to enquire whether he would pay the remainder of her bills. Rockwell agreed and paid the final bill. Nevertheless, Gentry was still crippled and without an income.

In an effort to assist Gentry with her finances, fliers at Roosevelt Field set aside November 2 as "Viola Gentry Day," a day on which special aerial exhibitions would be performed for the general public. The total amount of money received in ticket sales was to be presented to Gentry. The event, however, was not a success, and only a small amount of money was raised.

Because of her financial and convalescent state, Gentry became a charity patient at the Hospital for the Ruptured and Crippled in Manhattan, where she remained for eighteen months. Physicians at the hospital reported that Gentry spoke constantly about her future aviation

"Viola Gentry—1929." This photograph might have been taken in October, when the Nassau County Hospital permitted Gentry a short car ride with friends. *Courtesy of the Elmo N. Pickerill Papers, Manuscript Division, Library of Congress, Washington, D.C.*

endeavors and agreed there was no medical reason for her not to fly again. Her doctors also noted it would be several months before she could care for herself. Despite that fact, Gentry was occasionally permitted to leave the hospital in the company of a nurse.

The timing of Gentry's transfer from Nassau County Hospital to the Hospital for the Ruptured and Crippled could not have been better. Notices had been mailed for all female pilots (one hundred American women were licensed to fly at that time) to attend a meeting on Saturday, November 2, 1929, in a hangar at Curtiss Field. Twenty-six women—including Gentry, who was presented a large bouquet of chrysanthemums upon her arrival—attended the historic meeting in which the Ninety-Nines, an international organization of female pilots, was formed.

At year's end, Gentry had a lot to look back on and much to be thankful for. She had flown into 1929 with great triumph and fanfare; she had faced tragedy and prevailed. On Christmas Eve, Gentry hung a stocking on the foot of her hospital bed. When she awoke on Christmas Day, Gentry's stocking, as told in *Hangar Flying*, was "bulging with everything from a Mickey Mouse to postage stamps and money, and the usual orange, apple, nuts, and peppermint candy."

First meeting of the Ninety-Nines at Curtiss Field, Valley Stream, Long Island, New York, on November 2, 1929. *Back row, left to right*: Neva Paris, Mary Alexander, Betty Huyler, Opal Logan Kunz, Jean David Hoyt, Jessie Keith-Miller, Amelia M. Earhart, Marjorie May Lesser, Sylvia A. Nelson, Dorothea Leh, Margaret F. O'Mara, Margery Brown, Mary Goodrich, Irene Chassey, "Keet" Mathews, E. Ruth Webb and Fay Gillis. *Front row, left to right*: Viola Gentry, Cecil "Teddy" Kenyon, Wilma L. Walsh, Frances Harrell and Meta Rotholz. *Courtesy of the International Women's Air & Space Museum, Cleveland, Ohio.*

the ninety nines inc.
International Women Pilots
NON-EXPIRING
VIOLA GENTRY
CHARTER MEMBER
Will Rogers Airport, Oklahoma City, OK 73159

Viola Gentry's non-expiring Ninety-Nines membership card. *Courtesy of Helen H. Codling.*

Now more than ever, Gentry believed her Christmas wish would come true. She knew she would fly again and looked forward to the coming year.

5
Regaining Altitude

1930-39

You can't keep Viola Gentry down.
—*"Viola Gentry Looks Up,"* Buffalo Courier Express, *July 12, 1930*

Gentry did not have to wait long for her Christmas wish to be granted. On December 31, 1929, she was allowed to leave the hospital for a flight with pilot Daniel J. Brimm. Newspapers across the country posted photographs of Gentry underneath headlines such as "Viola Gentry Flies Again" and "Viola Gentry, Out Again, Up Again."

Readers were no doubt interested to read of the woman who, after being given a one in three chance to live, had beaten the odds and taken to the sky once more. An undocumented newspaper clipping in the collection of the Ninety-Nines, Incorporated, titled "Disabled in Crash, Aviatrix Flies Again" and dated January 2, 1930, revealed:

> *ROOSEVELT FIELD, N.Y., Jan. 2—(INS)—Miss Viola Gentry, courageous woman flyer whose nerve apparently has not been harmed by the serious injuries she suffered in an airplane accident...rode in an airplane Tuesday the first time since the crash...*
>
> *Miss Gentry has been a patient in a New York hospital since her release from Nassau Hospital...Her right arm is still of little use, and has to be carried in a sling, but she is able to use her left arm...She had looked forward eagerly for a long time to the moment when she could go into the air again and she seemed very happy when she got into the plane and when she left it...*

> *"It was like heaven," she said after her arrival here. "I promised the doctor I would take care of myself and so I could not get up near the controls. I was a little cold, but I didn't mind that."*
>
> *"Sure I am," said Gentry when she was asked whether she was going to continue in aviation. "I have only one neck to break and that's not broken yet…My left arm is all right, but my right arm is still in a cast. If it is healing when they take off the cast in a few weeks, everything will be all right. If not, it means another operation. It feels good to get back with the old crowd here again."*
>
> *More than thirty pilots, mechanics and other employes* [sic] *who have known Miss Gentry for a long time and admire her for her pluck, shook hands with her and congratulated her on the improvement in her condition.*

Gentry's pilot's license was scheduled to expire on February 28, but because of the condition of her arms, she was unable to renew it. Worried she would lose her initial license number, 1822, Gentry asked a friend to write a letter to Clarence M. Young at the Department of Commerce. On February 14, 1930, [Martha] E. Goldschmidt composed a letter to Young, which stated:

> *Dear Mr. Young:*
>
> *This letter is written by a friend of Miss Viola Gentry, because she has not as yet recovered the use of her arms since her accident…*
>
> *On February 28th her license expires, and she wishes to know whether or not some special arrangement can be made to keep it in effect. The number is: 1822*
>
> *If the license is permitted to expire she would have difficulty for awhile in qualifying for another one. She is able to fly a plane but cannot take off or land it. Her right shoulder is mending slowly and she still carries her right arm in a cumbersome metal brace. The left arm is responding to treatment, but as yet does not function normally…*
>
> *She has asked me to thank you for the many courtesies you have shown her and to assure you that if you are to aid her in this situation she will be deeply grateful.*
>
> *Sincerely yours,*
> [Martha] *E. Goldschmidt*

A few days later, Young responded:

> *My dear Miss Gentry:*
>
> *With reference to Miss Goldschmidt's letter…relative to the receiving of your private pilot's license, I am sorry to inform you that it will be necessary*

for you to submit a new medical examination before you can resume flying under your present license.

We feel that although we could temporarily renew your license now, it would be of no use to you until such time as you had submitted a new medical examination. I would suggest that you return your license to this office, in which case we will hold your number, 1822, open so that you may be reissued this same number at such time as your license is renewed.

With kindest regard and best wishes for your speedy recovery, I am

Very sincerely,
Clarence Young

Both letters, which are in the files of the Federal Aviation Administration, document Gentry's steadfast belief that she would heal and once again assume control of an airplane. In regard to her license, Gentry took Young's advice and turned it in. From that point on, she concentrated on her health and waited for the day her license would be reinstated.

Throughout her stay at the Hospital for the Ruptured and Crippled, Gentry remained positive even when circumstances worked to discourage her. She was blessed with a large, and loyal, circle of friends—companions who stood by her, lifted her spirits and assisted with her expenses.

On March 29, 1930, the First Annual Aviation Ball, given in honor of Gentry and Charles DeBever, was held at Roosevelt Field. DeBever, an aviation stunt man and parachutist, had been wounded during a parachute jump. The ball, which cost $1 per person, featured music by the Meyer Davis Orchestra and performances from a variety of Broadway headliners.

First Annual Aviation Ball ribbon worn by Viola Gentry on March 29, 1930. *Courtesy of the International Women's Air & Space Museum, Cleveland, Ohio.*

The event was a success, and at its completion, Gentry and DeBever were each presented $765.

Gentry appreciated the kindness of her friends and never failed to repay them, in one way or another, for all they had done. In November, on one of her weekend outings from the hospital, Gentry traveled to Middletown, New York, to visit Wallace "Wally" Bishop. She was grateful he had heeded the call from the Nassau Hospital when she needed blood and believed his donation saved her arm from being amputated. For that, she wanted to thank him in person. The *Middletown Times Herald* told of Gentry's visit, gave an update on her physical condition and revealed her future aviation aims. The article "Girl Flier in Chester, Traces Comic Artist Who Saved Life After Crash," which was published on November 24, 1930, said:

> *Although she is suffering still from the injury to her arms, Miss Gentry has not lost any of her interest in aviation. She cannot handle a plane entirely alone because of the limited action of her arms, but she has flown much with others and has handled planes in the air recently, she said.*
>
> *Within a few months she hopes to be able to attempt something that will give her the claim to fame she feels she lost when fog and lack of fuel put a disastrous end to her endurance flight.*
>
> *Afraid to go up again? Afraid to attempt greater things? Oh, no. Not Viola Gentry.*
>
> *"I was saved for something…and I'm going to take up flying where I left off at the time of the accident…"*
>
> *Meanwhile she is preparing to fight fog in the future by knowledge of radio. She is a student of Randy* [Enslow] *radio operator…of New Haven. Before she flies her own plane again she will be a licensed radio operator also.*

Although still under the care of a doctor and void of a job or income, Gentry worked to raise funds for those in need. In December, Gentry used her celebrity status to help sell benefit tickets for a dinner, a dance and airplane rides over New York City. Gentry, along with other famous female pilots, joined with the Women's Emergency Aid Committee for Unemployment in an effort to raise $500,000 for the Salvation Army's "Relief for the Unemployed" campaign. As reported in the article "Women Fliers Aid Relief Fund Drive," which was published in the *New York Times* on December 16, 1930, there were fifty thousand unemployed "heads of families" in New York City. The unemployment situation was so dire that "Lieut. Col. Edward Underwood declared that if the Salvation Army had $1,000,000 for immediate use, it would not meet the present demands for relief."

Whether the women met their goal is unknown. However, the economy continued to decline.

At the end of the year, Gentry was still a patient at the Hospital for the Ruptured and Crippled. She had spent two Christmases at the facility, but by the spring of 1931, she resided there no more. After her release, according to *Hangar Flying*, Gentry took up residence with Mrs. Thomas Grogan, a nurse, as she still required assistance with daily activities.

Viola Gentry, early 1930s. *Courtesy of the author.*

The *Winnie Mae*, August 31, 1933. *Courtesy of the author.*

By mid-year, Gentry was spending the majority of her time at Roosevelt Field. "If I couldn't fly," she said in *Hangar Flying*, "I could at least watch those who could."

During the summer of 1931, fliers Wiley Post and Harold Gatty arrived at Roosevelt Field to prepare for their around-the-world flight. As reported in the article "Post and Gatty to Take Wives with Them on Trip to Oklahoma in Their World-Girdling Plane," which was published in the *Evening Independent* on July 3, Gentry "was the first to befriend" the two fliers upon their arrival.

While in New York, Post and Gatty attended various functions with Gentry and a host of other pilots. On one occasion, Post took Gentry and her friend Sylvia Smith for a short flight. Whether the flight occurred in the *Winnie Mae*, Post and Gatty's Lockheed Vega, is unknown, but it is a reasonable assumption.

On June 23, Post and Gatty took flight from Roosevelt Field in an attempt to fly around the world. After eight days, the fliers completed their mission and returned to a jubilant homecoming. According to the abovementioned article, "While they were circling the globe [Gentry] took care of the clothes they left behind, darning their socks and having their linen laundered so that they would have a change waiting for them when they got back. She was on the field when they completed their world flight and was at Mrs. Post's side all during the nervous hour before the landing. Today the fliers greeted her as an old friend."

Post and Gatty were instant celebrities. The fanfare surrounding their flight re-energized Gentry's desire to set another aviation record herself. Yet in order to do so, she needed a pilot's license, an airplane and the appropriate funds—none of which she possessed.

By late summer, Gentry was eager to obtain employment, find lodging of her own and return to her pre-crash way of life. Floyd Bennett Field, New York's City's first municipal airport, had been dedicated in May, 1931, and it was there Gentry first sought employment. She hoped the new airfield would need a hostess, but according to a letter, which is in the collection of the International Women's Air & Space Museum, addressed to her from the commissioner of docks, John McKenzie, there was "no such position as 'Airport Hostess' provided in the personnel for operation of the Floyd Bennett Air Field."

Gentry's efforts to acquire a job in the field of aviation proved futile. As a result, she found it necessary to return to her former occupation. The *Utica Daily Press*, on September 24, 1931, announced, "Viola Gentry, who gained some reputation as an aviator, has recovered from her latest crash and has gone back to her old job as a cashier in a restaurant in Brooklyn. Flying may be romantic, but it is also dangerous…Besides, it is nice to have the well known pay envelope and that is not always forthcoming in the flying game."

The *Bee*, on September 21, in the article "Viola Gentry Has Given Up Aviation Now," stated, "Physical disability does not appear to be the main reason for her withdrawal from aviation. 'Depression has hit aviation as it has hit everything else,' she said."

The *Statesville Record & Landmark*, on September 29, further declared, "Viola Gentry…is back on her little job in a restaurant and 'glad to get it.' She flew far, wide and handsome for a time but learned that fame does not always spell financial independence and so came back to an humbler but a surer means of livelihood."

Gentry became employed as a cashier at Pierre's, an upscale French restaurant in Brooklyn, on September 21. There, she worked, according to *Hangar Flying*, from 11:30 a.m. until 8:30 p.m. six days a week. On her days off and in the early morning, she spent time at Roosevelt or Floyd Bennett Field. It was a routine similar to her pre-1928 record-setting flight.

While working at Pierre's, Gentry treated Post, Gatty and five others to dinner. As the dinner occurred during her work shift, Gentry hosted her guests and attended the cash register at the same time. The *Brooklyn Daily Eagle* gave an account of the gathering on October 1. The article "Viola Gentry Stands Post and Gatty 'Treat'" said:

Postcard of Floyd Bennett Field, Brooklyn, New York, circa 1931. *Courtesy of the author.*

> *Last night with some friends,* [Post and Gatty] *took their first flight to Brooklyn and made a safe landing—from a taxi.*
>
> *Pierre's Restaurant…was their destination. They were greeted by a welcoming committee of one—Viola Gentry, famous woman flyer.*
>
> *Miss Gentry did not present the distinguished aviators with the key to the borough. But she did give them and her other five guests a prize dinner.*
>
> *Heavy traffic…delayed their scheduled 7 p.m. arrival…their hostess…kept a lookout from…behind the cash register.*
>
> *It's there the aviatrix performs stunts these days. Punching the keys and handing out change with a cheery "Thank you" comprise her program.*
>
> *Not breathtaking, she admits matter-of-factly, but guaranteed to bring her a weekly wage for the Winter without any broken bones…*
>
> *Hostess Gentry divided honors with Cashier Gentry. Every so often Miss Gentry would leave her guests to take care of a regular patron. "Business must go on" is her motto.*

As Gentry listened to Post and Gatty, her mind must have turned inward, where scenes and emotions from her own record-setting flight were revealed. She was more than ready to fly again, to take the controls and to set another record.

Within weeks of the dinner, Gentry had completed the forms to renew her pilot's license and undergone a physical examination. Afterward, she waited for the Department of Commerce's decision. On November 4, 1931,

Gentry received a letter, a copy of which is in the files of the Federal Aviation Administration, from Gilbert G. Budwig, director of air regulation. Budwig told Gentry she did not meet the physical requirements necessary for a pilot's license, and her renewal had therefore been denied.

Gentry did not give up. Instead, she flew to Washington, D.C., and submitted an application to Clarence M. Young for a waiver that would permit her to renew her license and fly alone. Her request, however, was not approved. Nevertheless, Gentry continued to fly with friends in order to log flight time and waited on edge for her right shoulder and arms to regain full mobility.

In January 1932, Gentry traveled to Miami, Florida, to attend the All-American Air Races. When she arrived, she discovered there were no air contests for women and decided to do something about it. The *Reno Evening Gazette*, on January 11, in the article "Women Fliers Demand Rights," said:

> *Miami, Fla. Jan 11—(AP)—Women fliers attending the all-American air meet here, displeased at failure of officials to provide contests for them*

Female pilots at the All-American Air Races in Miami, Florida, January 1932. *Left to right*: Mrs. John T. Remey, Edith Descomb, Mrs. Freddie Lund (Bettie), Helen G. Fitzgerald, Jean Lenore Stiles, Smaranda Brăescu (Romanian parachutist), Viola Gentry and Mrs. Robert Moore. *Courtesy of the State Archives of Florida, Florida Memory.*

> *issued a joint demand today that they be allowed to participate in the remaining part of the program.*
>
> *The group was led by Viola Gentry, New York flier, and Mrs.* [Bettie] *Lund, widow of the famous stunt flier.*
>
> *As a result of the complaint, race officials arranged for three events for women pilots.*

Although Gentry was unable to compete, she was ready and able to take up the cause for women's rights in aviation. In fact, she took every opportunity to encourage women to fly. In late February or March 1932, Gentry received a note from a woman asking for an autographed picture. Her letter of response, which is in a private collection, stated:

> *Dear Frances Norris,—*
>
> *I will be very glad to send you an autographed photo, when I get some new ones which will be soon—*
>
> *Why don't you fly yourself? We girls have to carry on you know—*
>
> *Many thanks for writing me, I will be happy to hear from you again*
>
> *sincerely,*
>
> *Viola Gentry*

Not long after her return to New York, Gentry took flight in a Monocoupe 70 with her friend John R. "Jack" Warren, who was an engineer and ensign in the Naval Reserve Unit at Floyd Bennett Field. Warren had been a guest of Gentry's at the Post and Gatty party and was referred to as an expert on spark plugs.

The two had flown from Floyd Bennett Field to Roosevelt Field. At the time of their landing, a gust of wind overturned the airplane. Newspaper reporters, and perhaps Gentry herself, reflected on the "jinx" of 1929 and wondered if it had pursued her into the '30s. An undocumented newspaper clipping in the collection of the International Women's Air & Space Museum told the story of Gentry's crash under the headline "Her Luck Holds."

The article stated:

> *Miss Viola Gentry is taking things easy in her home…today, reflecting on the good luck of her bad luck.*
>
> *The girl flyer…was in another accident yesterday when the plane in which she was flying was upset by a gust of wind as it landed at Roosevelt*

John R. "Jack" Warren, early 1930s. *Courtesy of Helen H. Codling.*

> *Field. Miss Gentry was not hurt and the pilot of the plane, John Warren, 28 of Belleville, N. J., also escaped injury…*
>
> *Their plane was wrecked in the crash. The two flyers were strapped to their seats and held fast in the wreckage. They were extracted by airport officers.*

In regard to Warren, as told by Helen H. Codling in an oral interview, Gentry "was crazy about [him]."

Warren obviously felt the same way about Gentry, as the two were later married. Their wedding, however, was a secret. An undocumented newspaper clipping, which is on file at the Cradle of Aviation Museum, proclaimed, "Viola Gentry Wed Secretly for Year to Jack Warren," and revealed:

> *Viola Gentry, famous aviatrix and Jack Warren…have been secretly married for nearly a year, they admitted last night at their home* [in] *Brooklyn.*
>
> *Only a few close friends and relatives, they said, were present at the ceremony. They refused to say when and where they were married and would not discuss their reasons for keeping the marriage a secret. The news leaked out yesterday at Roosevelt Field.*

Over forty years later, Gentry shared the reason for her secret wedding with George Vecsey. In G*etting Off the Ground*, Vecsey documented Gentry as saying, "[Jack's] family thought I was a disgrace to women because of my flying…His mother backed me up, but that was all."

Although likely disheartened by the opinions of Warren's family, Gentry did not cease her efforts to promote women in aviation or regain her pilot's license. In fact, Gentry sent several additional appeals to the Department of Commerce regarding the renewal of her license. On June 7, 1932, Inspector Oren P. Harwood of the Department of Commerce agreed to give Gentry a special flying test to determine whether she should be allowed to pilot an airplane again. Gentry passed Harwood's test and at long last was in possession of a pilot's license.

The *New York World-Telegram*, on June 10, published Geraldine Sartain's article "9 Months in Cast, Then to the Clouds—'It's Heaven,' Sighs Flying Viola Gentry." In it, Sartain wrote:

> *The little girl from North Carolina who refused to be beaten at the flying game by an injured arm that will probably always be stiff gives a boyish grin and sighs, "Oh, it was heaven to be back flying a ship again by myself."*
>
> *...Miss Gentry still thinks she...may fly the Atlantic some time, but concedes it's only a vague dream. Now that the Department of Commerce waived her disability, gave her a special test three days ago, and granted her a new pilot's license, she's content for the while to make short flights here and there.*
>
> *What she really wants to do more than anything else in the world is to be given a job by some commercial flying outfit.*
>
> *"I don't want to go back to restaurant cashiering...Not that it wasn't worth it...But it's been three long years between times when I had hold of the stick of a plane, and I want to give all my time to aviation now."*

Because Gentry was married and did not have to earn a living on her own, she was able to leave Pierre's and focus on aviation. Gentry was often called on to give speeches on aviation, officiate at air races and attend various other aviation-related functions. When Amelia Earhart returned to New York on June 20, after flying solo across the Atlantic Ocean, Gentry accompanied—at the request of Mayor James J. Walker of New York—the official group that greeted her. She was also sent an invitation to Earhart's official reception at city hall, which she attended.

Although Gentry never lost her desire to compete in another refueling endurance flight, Earhart's success, in addition to Post and Gatty's, likely added fuel to the flame. In late June, Gentry spoke with Charles S. "Casey" Jones about arranging such a flight, the details of which were recorded in *Hangar Flying*:

> *We agreed that while I could arrange such a flight, it would be impossible for me to fly in it. My license had just been restored but it carried a "waiver"*

Viola Gentry (seen right in light-colored hat and dress) was part of the official group that met Amelia Earhart when she returned to New York on June 20, 1932, after her solo flight across the Atlantic Ocean. *Courtesy of the International Women's Air & Space Museum, Cleveland, Ohio.*

with it, and it was felt it would be too dangerous for me to actually fly in such an event. Casey agreed to supply the plane for the flight, as well as the refueling plane. The plane used would be a Curtiss Thrush with a 240 hp. Wright Whirlwind motor. At that time, the women's refueling record was only five days and Casey thought, with any kind of a break, a new record could be easily established.

Jones wrote to Frances Harrell Marsalis, who worked for the Curtiss-Exhibition Team, and asked if she and Louise M. Thaden, a record-setting pilot and winner of the first Women's Air Derby, would fly the plane. Both women agreed. Gentry managed the flight and reported the outcome in *Hangar Flying*:

Everything went off according to schedule. The Curtiss Thrush was named The Flying Boudoir, and was sponsored by the I.J. Fox Fur Company. The flight began on Sunday, August 14, 1932, at 2 P.M., and came back on earth on Monday, August 22, at 6:05 P.M., having been in the air for 8 days, 4 hours, 5 minutes and 48 seconds—or roughly 196 hours.

Left to right: Viola Gentry, Louise M. Thaden, Frances Harrell Marsalis and Charles S. "Casey" Jones stand in front of *The Flying Boudoir*, August 1932. *Courtesy of the International Women's Air & Space Museum, Cleveland, Ohio.*

> *The flight was a good attention-getter—but a financial flop. The girls had a series of special appearances lined up after the flight was over, but they just about broke even.*

Although Marsalis and Thaden had set a new record, there was little gained by it. Russell Plehinger, in his book *Marathon Flyers*, stated, "Only a small crowd had gathered to see ladybirds Louise Thaden and Frances Marsalis alight after a record eight days aloft over the Big Apple. Flight manager, Viola Gentry…and field manager Casey Jones were on hand to welcome them but the novelty of endurance flying was beginning to wear off."

Despite the seemingly lack of interest in endurance flights, Gentry and her husband decided they would sponsor and market them. In addition to their endurance flight business, Gentry designed a clothing line, Viola Gentry Skyline Fashions, which consisted of divided skirts made from flannel and worsted. She had hoped it would be a profitable business, as it offered female fliers feminine, flight-friendly attire. Marsalis and Thaden wore Gentry's skirts on their record-breaking flight and, according to

Hangar Flying, proclaimed them to be "very satisfactory—not binding or tight, but comfortably loose."

Unfortunately, Gentry's fashion line was a "financial flop," too. The manufacturer, according to Gentry in *Hangar Flying*, said, "It was a hopeless job to try to sell [the divided skirts]—and therefore, gave up the line."

The following year, Gentry staged her own comeback in the world of aviation. She regained her sporting license and entered the Second Annual Annette Gipson All-Women Air Race. Gipson, a pilot who had wanted to compete in a women's air race in 1932, discovered none had been scheduled on the East Coast, so she sponsored one herself.

The Second Annual Annette Gipson All-Women Air Race was held on June 4, 1933, at Floyd Bennett Field. The race consisted of a forty-mile, two-lap

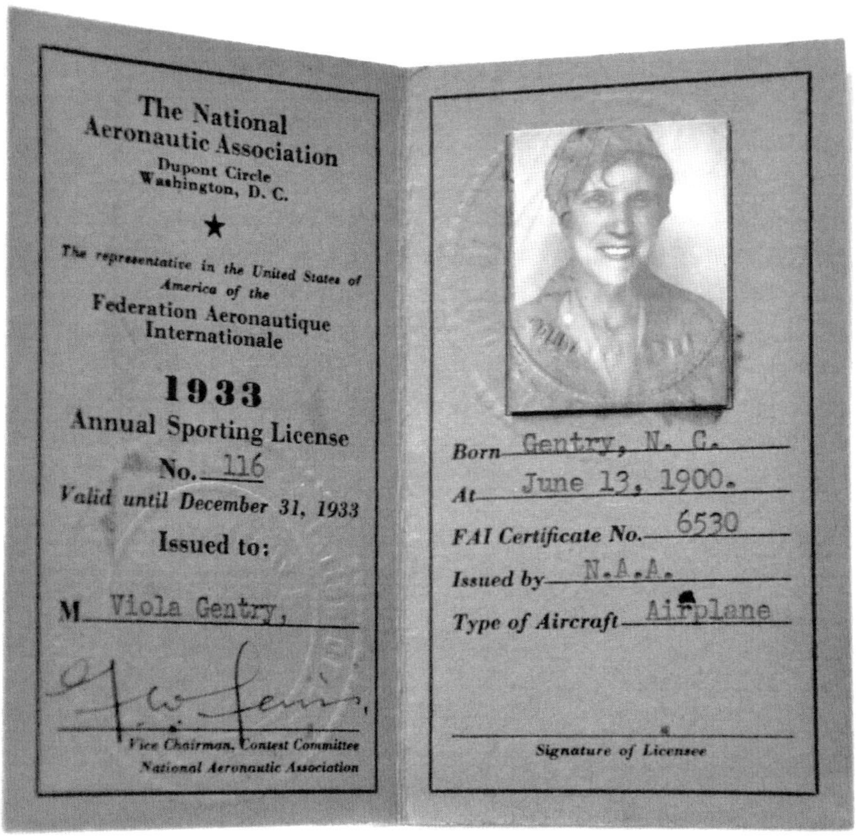

The National Aeronautic Association
Dupont Circle
Washington, D. C.

★

The representative in the United States of America of the
Federation Aeronautique Internationale

1933
Annual Sporting License
No. 116
Valid until December 31, 1933
Issued to:
M Viola Gentry,

Vice Chairman, Contest Committee
National Aeronautic Association

Born Gentry, N. C.
At June 13, 1900.
FAI Certificate No. 6530
Issued by N.A.A.
Type of Aircraft Airplane

Signature of Licensee

Viola Gentry's Fédération Aéronautique Internationale/National Aeronautic Association of the United States annual sporting license, issued in 1933. *Courtesy of the International Women's Air & Space Museum, Cleveland, Ohio.*

Annette Gipson trade card, early to mid-1930s. *Courtesy of the International Women's Air & Space Museum, Cleveland, Ohio.*

Viola Gentry raced a Waco airplane in the Second Annual Annette Gipson All-Women Air Race on June 4, 1933. *Courtesy of the International Women's Air & Space Museum, Cleveland, Ohio.*

course between Floyd Bennett Field and Curtiss Field. Twenty-three women, including Gentry, registered for the race. Gentry said in *Hangar Flying*, "I knew I could not possibly win, but at least I could try."

Gentry finished last but said, "I was not at all disappointed."

At the end of the race, entrants were treated to dinner at the Half Moon Hotel on Coney Island. Awards and speeches were given during the event, and to Gentry's surprise, she was honored. Earl Southey, a World War I pilot, presented her with "a special trophy—a plaster replica of the Birdman Trophy which had been given to Lindbergh."

Of the occasion, Gentry said, "I know I will never forget the words of Earl Southey when he presented the trophy—'To the most courageous girl in aviation—one who has—I don't care if it is a vulgar term—plenty of guts.'"

Gentry had overcome her wounds, fought to regain her pilot's license and, in 1933, conquered the clouds once more. Her mind was sky bound, and she reveled in every opportunity to take flight.

On September 11, Gentry took part in President Franklin D. Roosevelt's National Recovery Administration (NRA) campaign. She, along with nine other female pilots, flew over New York City and, when they reached Fifth Avenue, dropped bouquets of carnations tied together with red, white and blue ribbons, all of which bore the insignia of the NRA. Their flight was the start of a three-day NRA demonstration that culminated on September 13 with a parade of 250,000 NRA supporters.

Yet flights such as those only increased Gentry's desire to attempt another refueling endurance flight. As a result, Gentry made arrangements with

Left to right: Viola Gentry, Mary Sansom and Roger Q. Williams, 1933. *Courtesy of the International Women's Air & Space Museum, Cleveland, Ohio.*

Mary Sansom, who held the New England record for continuous loops in an airplane by a woman, to attempt a flight with her in October. Financed by Gentry's husband, the airplane, which was a Curtiss Thrush with a 225-horsepower Wright motor, was named *Outdoor Girl*, as they had received sponsorship from Outdoor Girl Cosmetics.

Newspapers ran stories of the upcoming flight, accompanied by photographs of the women and their airplane. Writers, who in many cases were women themselves, gave brief biographies of Gentry and Sansom. They also provided physical descriptions of the women, including details such as hair and eye color; one writer pointed out that the women's blue eyes matched their flight suits. Descriptions of the women's food selections, remarks about knitting in their spare time and talk of special makeup onboard equated the flight to a flying slumber party. More important facts, such as the physical and mental stamina needed to complete an endurance flight and the uncomfortable conditions within the airplane, were not often mentioned.

On October 16, Cora Bennett christened the *Outdoor Girl* with a bottle of champagne in front of numerous onlookers at Floyd Bennett Field.

Gentry and Sansom were scheduled to depart two days later. However, just before takeoff, Sansom determined she could not make the flight. The reason behind her decision is unknown, but there is no doubt Gentry was disheartened by the announcement.

Gentry asked Frances Harrell Marsalis, who had flown *The Flying Boudoir* and co-held the women's refueling endurance flight record, to assume Sansom's place. Marsalis agreed, and new plans were made. The women would fly the same airplane and have the same sponsor, but this time the flight would take place in Florida.

In Miami, Florida, on December 10, 1933, Gentry, Marsalis and the *Outdoor Girl* took flight. According to the article "Plane of Girls Is Downed," which was published in the *Evening News* on December 11, Gentry exclaimed just prior to takeoff, "See you in 1934."

After twenty-one hours in flight, however, high winds forced the women to land. A minor setback, Gentry and Marsalis rested and prepared to take off again on December 15. Yet the flight, at least with Gentry onboard, was not to be. The article "Viola Gentry Collapses in Endurance Hop Ship," which was published in the *Milwaukee Sentinel* on December 16, best describes the turn of events:

Cora Bennett christened the *Outdoor Girl* on October 16, 1933, at Floyd Bennett Field. Viola Gentry and Mary Sansom stand to the left in flight suits. *Courtesy of the International Women's Air & Space Museum, Cleveland, Ohio.*

Viola Gentry inside the *Outdoor Girl*, 1933. *Courtesy of the International Women's Air & Space Museum, Cleveland, Ohio.*

> *Miami Beach, Fla. Dec. 15 (Universal)—The tragic jinx that has dogged the flying trail of Viola Gentry struck again this afternoon when the gallant feminine flier collapsed at the controls as she was about to take off on her second Florida attempt to break the women's endurance record.*
>
> *The plane…was fuelled* [sic], *warmed up and waiting at the Municipal airport this afternoon.*
>
> *Miss Gentry was at the controls, and Mrs. Marsalis was ready to climb in…when suddenly the former crumpled unconscious in her seat.*
>
> *She was taken to University hospital here, where physicians say she was suffering from appendicitis. Late tonight it was reported at the hospital that Miss Gentry was not in immediate danger, and…if she regained enough strength…would be sent by plane to New York for an operation.*

Marsalis telephoned Helen Richey—the first licensed female pilot in Allegheny County, Pennsylvania, and first woman to acquire a commercial pilot's license—and asked her if she would replace Gentry.

Richey agreed, and on December 20, the two women began their flight. Nearly ten days later—after being in the air for 247 hours and 42 minutes—Marsalis and Richey landed. They had set a new refueling endurance flight record for women.

News of their success must have come as quite a thrill for Gentry. She cheered and celebrated with the women and, when she recovered from her appendicitis, made public appearances with them. On the outside, Gentry rejoiced in their success, but on the inside, she must have shed tears of disappointment.

The new year of 1934 did not offer Gentry or her husband much encouragement. On July 31, Gentry and Warren filed bankruptcy in federal court. They listed their debts as $1,651.26 and claimed no assets. The bankruptcy, according to the article "Viola Gentry Bankrupt," which appeared in the *New York Times* on August 1, 1934, was filed under the "co-partnership of Warren [and] Warren…and described its business as 'endurance flights.'"

Four days later, Gentry was further distressed when she learned that her friend Frances Harrell Marsalis had died. Marsalis had been flying in the Cleveland Air Races on August 5 when she lost control of her airplane and crashed. A memorial service for Marsalis was held at Roosevelt Field, where many noted female pilots, including Gentry, were in attendance.

It is no surprise that after sickness, missed opportunities, financial loss and the death of a friend, Gentry was an emotional wreck. The September 8, 1934 edition of Talbot Lake's *Daily News* column "Profiles for Today" told of Gentry's woes. The words "Viola Gentry, Charming Woman Flyer, Grounded Once Again—This Time by the Depression" appeared underneath the column title. Lake said:

> *The fortune of our romantic flying girls—the intrepid feminists who have conquered the air—seem to be having an ill period. Recently the tragic death of Mrs. Marsalis stunned the flying fraternity, and now we hear that the charming Viola Gentry…has become bankrupt. This is not like being killed, but nevertheless it is a drastically serious matter to the "flying cashier." She and her husband were not able to get their projected endurance flights into the air, and they seem to be grounded for an indefinite period.*
>
> *This is bitter medicine to one of so adventurous a nature as Miss Gentry, for her story is that of glamor and accomplishment from small beginnings…*
>
> *She attained national fame in 1928 when she stayed up more than eight hours, making what was then the women's endurance record. Her luck seemed to desert after this, for in June, 1929, she was badly hurt in a crash…She went up again, and appeared to be on the way to recover her*

> *lost laurels when appendicitis laid her low...she planned further endurance flights, but the depression caught up, and now Miss Gentry is grounded once more.*

For the remainder of the year and throughout the first half of 1935, Gentry spent time recovering physically, emotionally and financially. By mid-1935, Gentry had returned to the newspapers, but this time, reporters told of her activities with the Women's International Association of Aeronautics (WIAA) and composed stories of her life as a wife and famous flier.

In November 1935, Gentry and her husband moved to Orange, New Jersey. Warren worked as a project engineer for the Breeze Corporation, and the move allowed him to be closer to his office. The change of scenery seemed to abide well with Gentry, as throughout 1936, she gave lectures and made special appearances at several New Jersey venues. Her happiness, however, did not last.

As if Gentry had not conversed enough with tragedy, it visited once more in January 1937. In December 1936 or early in January 1937, Gentry's husband was afflicted with pneumonia and hospitalized. On Friday, January 15, John R. "Jack" Warren passed away. An account of his life, which was published in the *Newark Evening News* on January 16, 1937, and titled "John Warren, Flyer, Dies of Pneumonia," read:

> *Husband of Former Viola Gentry Was Engineer with Plane Concern*
>
> *John Warren, veteran flyer, and husband of the former Viola Gentry, one time holder of the endurance flying record for women, died yesterday in Presbyterian Hospital from pneumonia...*
>
> *One of the first pilots to fly at Floyd Bennett Field, Mr. Warren was an ensign in the Naval Reserve...*
>
> *In recent years, Mr. Warren restricted his flying to trips necessary in connection with his engineering work.*
>
> *He was a member of the Quiet Birdmen of New York and the Aircrafters of Philadelphia.*

Warren's funeral service was held in a hangar at Floyd Bennett Field. Afterward, Gentry took a somber flight with Lieutenant John Remmert, and in accordance with her husband's wish, she released his ashes over the field. As reported by the *Brooklyn Daily Eagle* in the short piece "Dead Flier's Ashes Scattered at Field" on January 20, 1937, "Three Navy ships [airplanes] in formation formed an escort and dropped red roses as a tribute."

In regard to Warren's death, Helen H. Codling said she remembered her aunt "sobbing when he died."

Gentry had endured much heartache, but nothing equaled the anguish of her husband's death.

As the year progressed, the decade continued to be a bearer of bad news. In June, Amelia Earhart and Fred Noonan took off from Miami in an attempt to fly around the world. Tragedy struck on July 2, 1937, when Earhart radioed she was running out of gas and was soon after not heard from again. The day after Earhart's disappearance, Gentry was documented in a Hammond, Indiana paper called the *Times* as saying, "Don't you worry about Amelia—10 to 1, and I'll bet on it, she'll come through okay…She's a grand girl and a grand aviatrix…You can't beat a combination like that. I know everything is going to be all right."

For nearly two weeks, search parties tried to locate Earhart and Noonan, but they were not to be found. Nevertheless, Gentry did not give up hope. Still blanketed by a weight of sorrow, she wanted to believe Earhart was alive. She needed something to hope for.

In November, Gentry and other members of the WIAA planned a tribute for Earhart and Noonan at Floyd Bennett Field. During the observance, which took place on November 21, speakers—such as Fanny Hurst, a popular novelist; Clara Adams, an aviation enthusiast; and Gentry—presented reasons they believed Earhart and Noonan were alive.

Throughout 1937, Gentry continued to stay active by speaking at and participating in aviation-related functions. In 1938, she produced a newsreel production titled *First Flights*, which featured footage of aviation's greatest flights. Dick Kirschbaum, who wrote the column "Air Lanes" for the *Newark Evening News*, documented the film on September 2. He stated:

> *The idea of making the pictorial record of aviation is Miss Gentry's. With her background of aviation and personal knowledge of most of the flights, she has been able to serve as editor of the picture. Her efforts have resulted in a smoothly running film of the events…*
>
> *During the running of the picture Miss Gentry or another aviation figure, will tell behind-the-scenes highlights in connection with each flight. The picture is made to run an hour. Several New Jersey schools already have booked the film as part of their visual educational program.*

No matter her circumstances, Gentry was dedicated to flight and remained involved in every aspect of it. Despite hardships and heartbreak, Gentry never ceased encouraging women to become involved in the growing field of aviation. As the 1930s came to close, the world was in turmoil, and war once again loomed on the horizon. Gentry looked toward the coming decade with hope and excitement. She saw women in new roles—duties that would involve aviation and the protection of their country.

6

Changing Course

1940-49

I consider [Viola Gentry] *one of the best informed women in aviation in the United States.*
—*Horace B. Wild in a letter to the Civil Service Commission, January 26, 1940*

By the end of 1939, attitudes regarding women in aviation had begun to fly in a positive direction. At that time, according to Claudia M. Oakes, in her book *United States Women in Aviation 1930–1939*, women "were finally being taken more seriously as good, professional pilots." Oakes further said:

> *Women were no longer oddities in any area of aviation. Be they stewardesses, engineers, businesswomen, or pilots, they had for the most part accomplished their goals of helping make air travel a standard means of transportation, and had proven to the world that women could be competent pilots.*

This fact is evidenced in a letter sent by the United States Civil Service Commission on December 26, 1939, to the Women's International Association of Aeronautics (WIAA), New York branch that announced the acceptance of applications for the position of special agent or assistant under direction of the Civil Aeronautics Authority. The letter, which is in the collection of the International Women's Air & Space Museum, also requested the notice be forwarded to any "persons who would probably be qualified and who would be interested in Federal employment."

Although the exact duties of a special agent or assistant are unknown, a person in that position likely helped enforce aviation regulations.

Mrs. Jessie R. Chamberlin, president of the WIAA, returned the letter with a note of recommendation and a list of qualifications for Viola Gentry. In addition, Chamberlin sent news of the position to Gentry, who, without haste, mailed the commission a letter of application, a résumé and two photographs of herself.

Gentry's résumé, which is also in the collection of the International Women's Air & Space Museum, listed her "Qualifications and Experiences" and read much like a memoir. She told the commission of her birth in North Carolina; the death of her mother; her first flight; the recent death of her husband; the airplanes she had flown; her 1929 crash and subsequent stay in the hospital; her current aviation endeavors; and her experience as a mechanic.

In regard to mechanical aptitude, Gentry revealed that in the summer of 1925, she had been employed as an apprentice mechanic at the "Curtiss Factory in Nassau, Long Island, New York." There, she worked for a year "preparing radiators, building gas tanks, constructing and covering wings in the dope room [room where fabric was treated] and on the assembly line."

Gentry further declared she could "tear down, reassemble and properly time any aircraft motor" and that she "thoroughly [understood] the details of motor insulation in transport ships."

For Gentry, it was imperative the commission understand she was not, and had never been, just a pretty face in the cockpit. She was a skilled pilot "in possession of a thorough knowledge, both practical and theoretical, of the field of aviation mechanics."

In her concluding remarks, Gentry expressed satisfaction with the commission for allowing women to apply for the aviation-related position, as well as her desire to be the one to acquire it. She wrote:

> *I have always cherished the idea and hope that women would take their proper place in aviation development and that women would contribute their full share in aviation progress. Now that the Civil Service Commission has made it possible for women to take the examination for Special Agent or Assistant, I believe that my qualifications expressed in the foregoing paragraphs will merit careful consideration by your honorable body.*

To validate her statements, Gentry asked Captain Horace B. Wild—a pilot who had flown prior to 1916 and was the first field director of Cicero

Field in Cicero, Illinois—to send the commission a letter of recommendation on her behalf. Upon receiving her request, Captain Wild wrote a letter to the commission. Dated January 26, 1940, the letter read:

> *Gentlemen:*
>
> *Viola Gentry has…requested that I write you gentlemen a character reference. It is with great pleasure that I submit the following:*
>
> *I have known Viola Gentry since 1925 and have been closely associated with her in aviation activities and organization work.*
>
> *I have watched with great interest the progress she has made and I consider her one of the best informed women in aviation in the United States.*
>
> *She has been highly honored on several occasions by outstanding organizations in America.*
>
> *Her accomplishments in the aviation field are a matter of record. She possesses a delightful, magnetic personality and that all-important qualification of being a good listener.*
>
> *In recommending Viola Gentry to your commission for appointment to the Civil Aeronautic Authority as Special Agent I feel sure that you will add a very efficient and capable member to that organization.*
>
> *Very truly yours,*
> *Horace B. Wild*

There is no doubt Gentry longed for a career that would foster and support her flying endeavors—a profession that would offer steady and full-time work in the field she loved. However, despite her qualifications and recommendations, there is no evidence Gentry acquired the position. Without question, Gentry believed she had a favorable chance at being hired for the position and was disappointed she was not. Nevertheless, just as she had in the past, Gentry continued to work in various occupations, make public appearances, log flight time, maintain her pilot's license and fulfill her duties to the WIAA, all while seeking out new aviation-related opportunities.

On June 28, 1940, Gentry submitted another application, but this time, it was for a Social Security number. Although she was employed and referred to in aviation circles as Viola Gentry, she applied for her Social Security number under her married name—Viola Gentry Warren. At the time of her application, she resided in Brooklyn and was employed at the Hotel New Yorker.

When Gentry was not working at the hotel, she enjoyed planning and attending WIAA events, as well as other aviation-related gatherings. On

The Forum Council of Bergen County
INCORPORATED
238 MAIN STREET
HACKENSACK, N. J
TEL. HACKENSACK 2-8241

OFFICERS
Wing Tabor Wetmore
PRESIDENT
Henry Spindler
FIRST VICE-PRESIDENT
Frederick J. Butterworth
SECOND VICE-PRESIDENT
Eleanor Anderson
THIRD VICE-PRESIDENT
Dave Taggert
FOURTH VICE-PRESIDENT
Walter J. Martin
FIFTH VICE-PRESIDENT
Emma B. King
SECRETARY
Irvin G. Herman
TREASURER

ROCHELLE PARK - "Build the Forum Council with good speakers, attractive and diversified programs and keep interest in the meetings," is the slogan of President Wing Tabor Wetmore, recently elected. The Forum Council meets monthly in Hackensack.

The attendance at the last two meetings, Mr. Wetmore says, demonstrated the wisdom of his decision.

FROM
The Herald-News, Passaic, N. J.
Tuesday, November 19, 1940

Fri., Feb., 21 - 8:30 P.M.
~Coming~
~to~
~Hackensack Y.M.C.A.~
~Viola Gentry~
~Holder BIRDMAN TROPHY for Courage~
(Lindberg only other holder)
SUBJECT~ "Good Jobs in the Flying Business."
~Auspices~
The Forum Council of Bergen County
INCORPORATED

Tune In~
"Viola Gentry will broadcast her "talk"
on topic~ "Good Jobs in the Flying Business"~
from Station WBNX New York ~ Fri., Feb, 21 ~ at 1:45 P.M.
Auspices
The Minute Men of America
Incorporated
Wing Tabor Wetmore
Chairman

Flyer announcing Viola Gentry's talk "Good Jobs in the Flying Business," circa 1941. Although Gentry worked at restaurants and hotels to support herself, she was always in the know when it came to aviation-related job opportunities. *Courtesy of the International Women's Air & Space Museum, Cleveland, Ohio.*

December 3, 1940, the WIAA sponsored a fashion show to benefit the American Red Cross. Held at Roosevelt Field Inn, stewardesses and wives of pilots modeled the latest fashions. Gentry did not model; instead she assisted in directing ladies on and off the fashion runway.

Postcard of the Hotel New Yorker, New York City, New York, postmarked 1938. *Courtesy of the author.*

Not long after—according to the *Brooklyn Eagle* article "Viola Gentry Hostess to Composers," which was published on February 13, 1941—"Viola Gentry…gave a cocktail party [February 11]…at [the Roosevelt Hotel] in Manhattan, in honor of Rod Lehigh and Dick Kirschbaum, whose song, 'Wings to Victory,' was dedicated…to the 119th Observation Squadron, U.S. Army Air Corps and the 44th Division Aviation."

Lehigh, who composed the music for the song, and Kirschbaum, who wrote the words, were both aviation editors for the *Newark Evening News*. The event was attended by a large crowd and included "thirty officers of the 119th Observation Squadron [who] adopted the song for the air corps."

That same year, Kirschbaum published a book titled *Fifty Famous Flyers*, which included comic-like sketches of pilots. In his preface, Kirschbaum stated, "While this first edition does not carry all the famous flyers in the country, it has many of the best known pilots now flying the commercially chartered and many unchartered airways. I have the privilege of knowing them all personally through my work, which is conducting 'Air Lanes' in the *News*, the only daily column in the world dealing exclusively with aviation."

Gentry, as well as other female pilots, was included in Kirschbaum's book.

As in the past, when Gentry entertained guests—outside of WIAA, or other aviation-related organization's functions—she generally held the affair at the location in which she worked. Therefore, it is likely she was employed by the Roosevelt Hotel at the time of Lehigh and Kirschbaum's party. Two

Postcard of the Roosevelt Hotel, New York City, New York, postmarked 1941. *Courtesy of the author.*

months later, Gentry hosted another party at the Roosevelt Hotel, which further supports this assumption.

On Wednesday, March 26, 1941, Paramount Pictures, Incorporated, released *I Wanted Wings*, a movie about three United States Army Air Corps recruits and their love interests. The film included breathtaking aerial sequences, which won it a special effects Oscar. Because of the film's focus on aviation, Gentry was no doubt eager to view and promote it. Bosley Crowther's review of the movie in the March 27, 1941 issue of the *New York Times* stated, "The awe and majesty of flying…is imparted to the audience completely by the magnificence revealed upon the screen."

For certain, Gentry had always sought opportunities in which she could share—through words and airplane rides—the "awe and majesty" of flight with those who had yet to leave the ground. As *I Wanted Wings* was another means by which non-fliers could experience flight, Gentry was happy to endorse it. A notice in the March 29, 1941 issue of the *Brooklyn Eagle* read:

> *Up to Her Old Tricks—Viola Gentry spent a busy Saturday flying from Roosevelt Field to the capitals of five States to invite Governors McGrath of Rhode Island, Saltonstall of Massachusetts, Hurley of Connecticut, Edison of New Jersey and Lehman of New York to attend a preview of the new movie* I Wanted Wings. *Miss Gentry flew a* [Rearwin] *Skyranger. That evening she was hostess to members of the Women's International Association of Aeronautics at the preview in the Astor Theater, and later held a reception at* [the Roosevelt Hotel].

Whether Gentry was paid to promote the movie is unknown, but as the flights were reminiscent of her failed 1926 Philadelphia Sesquicentennial International Exposition plan, she likely devised, and benefitted, from the idea.

I Wanted Wings not only showcased the thrill of flight, but it also brought forth a sense of patriotism among its viewers. In his review, Crowther proclaimed the movie "a dependable inspiration to the youth of the land" and predicted it would "set things humming around recruiting and box offices."

In April 1941, the United States had not entered World War II but marched steadily toward it. Much like the patriotic speeches of World War I, movies-such as *I Wanted Wings* encouraged citizens to support and fund war-related efforts. When the Adelphi College Center for Long Island Women in Home Defense put forth an initiative to register 100,000 Nassau County women for a home defense unit, Gentry supported their cause. On May 3, the center held a daylong "Women's Home Defense Day" to educate women—by way of speakers, demonstrations and exhibitions—on the importance of creating a local defense program. According to the April 27, 1941 *Brooklyn Eagle* article "100,000 Women Sought for Home Defense Unit," one of the exhibits featured "radio and telegraphy equipment supervised by Viola Gentry, noted flier and communications expert."

The article further stated, "Convinced of the need for women as radio and telegraph operators in case of national emergency, Miss Gentry will explain how courses in the subjects may be established for girls."

On days she was not supporting war efforts, working at a hotel or promoting an event, Gentry flew for the fun of it and visited longtime friends. In May, Gentry traveled to Miami, Florida, to spend time with her friend Annette Gipson. While there, Maybelle Manning, a writer for the *Miami Daily News*, interviewed Gentry for her column "Talk of the Tower." Published on May 18, the article read:

> *Meet Miami's Illustrious Guest Flyer:—Viola Gentry, piquant, trim and violet-eyed and with the gayest and warmest of smiles. It was an enchanting visit we had with Viola out at the beautiful estate of Annette Gibson, herself a renowned flyer.*
>
> *"It's as much fun as visiting a rich child used to be," said Viola. "One who had two ponies and two basket carts, so that we could each ride and drive. Annette has three planes. And, oh boy, are we going to have fun!"*

Manning told readers of Gentry's early flying adventures and of her friendship with Amelia Earhart. In regard to Earhart, Gentry said, "Nobody ever had such an all inspiring way and understanding of people as Amelia. She was like a dancing sunbeam."

Gentry also shared her belief, just as she had after Earhart's initial disappearance in 1937, that Earhart was still alive. She did not want the hunt for Earhart to be abandoned and had plotted areas in which to search. Gentry said, "It is my duty and my sole life purpose to go and get Amelia…It is like an everspringing [sic] and never-satisfied intuition that makes me feel that she is waiting, somewhere alive."

At the end of the article, Manning said of Gentry, "I shall long remember my visit with Viola, her radiantly clear face, her arrestingly kind gaze with eyes of fearless inquisitiveness."

Two months later, on July 15, the North Miami Woman's Club presented the program "Women in Aviation." Gentry, who was the featured speaker, gave the presentation "Amelia Earhart Is Still Alive."

Viola Gentry was photographed upon her arrival in Miami, Florida, 1941. *Courtesy of the International Women's Air & Space Museum, Cleveland, Ohio.*

In her speech, Gentry revealed why she believed Earhart was alive. She also told of her plans to renew the hunt for Earhart. Around the globe, however, war raged, and the United States prepared for the inevitable. As such, Gentry's efforts to resume the search for Earhart were unsuccessful.

The impending war not only halted Gentry's attempt to locate Earhart but also placed her on a course in which she, too, would go missing or at least be absent from the media. On August 28, 1941, the *Mansfield News-Journal* reported that Gentry was still in Florida but would "soon begin ferrying bombers to England."

Without a doubt, Gentry would have been eager to participate in such an important venture. Her friend Jacqueline "Jackie" Cochran, a record-setting pilot and president of the Ninety-Nines, actively recruited female pilots for the British Air Transport Auxiliary, and it is possible Gentry answered her call. There is no evidence, however, to support the fact she ever ferried bombers. Although the organization often overlooked physical handicaps if a pilot could control the airplane, Gentry's limited mobility in her right shoulder and left elbow might have proved too much of a hindrance.

In October 1941, Gentry returned to New York from Miami and was welcomed home with a reception by the WIAA. Not long after, the United States entered World War II. Women, by way of home defense organizations and United States government posters, were reminded that their services were needed to win the war. Be it growing or preserving food, buying bonds, enlisting as a nurse or working full time in an industry, their wartime job was important and necessary.

World War II poster published by the Office of War Information, Washington, D.C., 1943. *Courtesy of the Prints and Photographs Division, Library of Congress, Washington, D.C.*

Gentry's mind likely churned with ideas on how her aviation knowledge and skills could benefit the war. However, from October 1941 until October 1942, there is little information regarding her activities or vocation. Two documents from January 1942—a *Brooklyn Eagle* article titled "Memorial Service Held for Former Navy Flier," which was published on January 16, and a Civil Aeronautics Administration form, which is in the files of the Federal Aviation Administration and dated January 21—revealed she held a memorial service at Floyd Bennett Field for her late husband, and her occupation was "lecturer."

On October 18, 1942, the following announcement appeared in the *Brooklyn Eagle*:

> *Woman Flier—Viola Gentry, one of the early women aviators, has left Brooklyn to take a position with an aircraft corporation in Kansas City. She was honored at a tea party at Schrafft's last week, by her colleagues in the New York Branch of the Women's International Association of Aeronautics. Miss Gentry is its vice president and Mrs. Jessie Chamberlain, president.*

Hal Wert, in his manuscript "Kansas City in World War II: Recovery and Transformation," which was composed for the Kansas City Museum's Community Curator program, remarked that aircraft manufacturers, along

Postcard showing "Union Station and Skyline, Kansas City, Missouri," circa 1940. *Courtesy of the author.*

with other war-related businesses, "provided employment for thousands." As a result, "From 1941–1945 the economy boomed, as people flocked to Kansas City for new jobs."

Gentry worked for one of those companies—Aircraft Accessories Corporation. Her exact position and duties at the corporation are unknown but would have involved the manufacture of communication equipment. An undocumented newspaper clipping on file at the Kansas City Public Library dated April 25, 1943, and titled "Bits of Crystal Ground Here Boost Fighting Fliers' Safety," said:

> *Inside* [Aircraft Accessories Corporation], *Kansas City's "white collar" assembly line workers (more than half of them women) ply ambidextrous fingers and war-acquired scientific knowledge to make what electrical engineers refer to poetically as "the beating heart of the radio transmitter."*
>
> *That "heart" is a postage-stamp sliver of quartz. It vibrates with the rapidity of radio frequency to carry the sound which guides army and navy aircraft safely on its way and steers it back home again.*
>
> *Crystal plant grinders, operators, inspectors, and testers virtually are aristocrats among war workers. Their training is intense, but the mental agility to grasp the intricacies of their work must be above average. Their task is complicated, but, it is not particularly dirty or heavy or dangerous. So the women, if they choose, can dress like anybody's stenographer—heeled pumps to white blouse—and the men can wear a business suit, minus the coat, with no harm either to garb or to efficiency.*
>
> *But, if you should ever get beyond the armed guard and the door mechanism warning which makes a sound like a sleepy goose, you would find nothing "white collar" in the aspect of the men and women busy inside the plant, in two shifts, twenty hours every day.*

Gentry was well suited for this type of work, as she had become interested in aircraft communication not long after she learned to fly. In fact, Gentry was a friend of Elmo N. Pickerill, who, according to her in *Hangar Flying*, became convinced in 1910 "that wireless signals could be transmitted to and from an airplane and…set out to prove it."

Pickerill did just that and, in 1920, "became associated with the Radio Corporation of America and was placed in charge of the Aeronautical Division to develop and manufacture radio apparatus of all kinds for the new air transport industry."

Gentry witnessed radio communication tests at Curtiss Field and remarked in *Hangar Flying*, "In the summer of 1928, a plane was being tested which had a radio installed. The plane was flown from Curtiss Field on Long Island to Dayton, Ohio, and during this flight the pilot was in constant touch with the ground by means of a two-way radio. I was one of the 'voices' privileged to talk to [Elmo] Pickerill (the pilot) from the ground."

In addition to her hands-on experience, Gentry understood how radio equipment could save lives. Had she and John W. "Big Jack" Ashcraft been able to communicate with other aircraft or authorities on the ground, their crash could have been avoided. In fact, Gentry so believed in the importance of radio communication that she became an avid student of the subject and, in 1931, was granted a radio operator's license, amateur class. Her intentions had been to operate a radio station at Roosevelt Field, but the plan did not mature.

Nevertheless, in Kansas City, Missouri, Gentry put her skills into action and used her knowledge for the greater good of the country. She had found, from the words of a United States government poster, a "war job" at a place where she "fit best."

On July 4, 1943, an advertisement for Aircraft Accessories Corporation was printed in the *Kansas City Star*. It read:

> *"Leader to Group…at Three o'clock…Bandits"*
>
> *The veteran squadron leader takes a quick squint into the sun—and a split second later—the pilot who might have been missing in action is warned by his seasoned commander. What a break that those fliers could talk together.*
>
> *Today communication equipment manufactured by Aircraft Accessories Corporation is helping Uncle Sam's fighters work together on land, sea, and in the air. The men and women on every shift in each A.A.C. factory feel the privilege and responsibility of what their work means to the safety of each American fighting man and to the total cause of victory. They work with such thoughts always in mind. And their work shows it.*

The words of the advertisement epitomized Gentry. No matter her role at the corporation, Gentry would have been gratified by her work, as it allowed her to satisfy wartime obligations and help save the lives of airmen. Helen H. Codling, in a written interview, said Gentry was "loyal to her [family]—loyal to longtime friends & flyers," and it is apparent she was loyal to her country, too.

Despite long hours at the corporation, Gentry managed to find time for love. It is unclear whether the flame of romance was ignited at work or if it had been lit prior to her arrival in Missouri, but it is documented that after three months in Kansas City, Gentry wed. On January 11, 1943, Gentry married Alexander McNight Cameron in the presence of a justice of the peace at the Jackson County, Missouri, Courthouse. On their wedding application, Cameron signed an affidavit stating that he was forty-one years of age, while Gentry stated she was forty-three. Her correct age was forty-eight.

Like Gentry, Cameron worked at Aircraft Accessories Corporation, so it is probable the two met at work. In regard to Cameron, there is little information regarding his life. Helen H. Codling, who lived in Kansas City during the time Gentry resided there, remembered Cameron as a "nice man" who worked as a "lathe operator."

Codling's first husband was a member of the United States Army Air Forces and trained in Kansas City. She accompanied him there and spent a considerable amount of time with her aunt and new uncle. In fact, Codling stated she often "cooked for them."

Gentry's time in Missouri must have proved happy. She had penned a new chapter in her life and was in the constant company of loved ones. Through correspondence, Gentry maintained friendships in New York and other locations across the country. On December 29, 1943, Gentry received a letter, which is in the collection of the International Women's Air & Space Museum, from Lady Grace Marguerite Hay Drummond-Hay, a British journalist and the first woman to travel around the world in an airship. The letter read:

> *My dear Viola,*
>
> *Please forgive me for not having written much sooner to offer my belated wishes for your Happiness. It was a great thrill to me, to learn of your marriage, and I am eager to meet your Husband. I want to be sure that he is good enough for you! That's what they all say in New York where you are so greatly beloved…*
>
> *With loving New Year greetings, yours*
>
> *Grace M. Hay Drummond-Hay*

Not long after, Drummond-Hay's sentiments rang true. The April 13, 1944 issue of the *Brooklyn Eagle* announced:

> *Alexander Camerons Honored at Dinner*
>
> *A dinner party was given Tuesday night at LaGuardia Field in honor of Mr. and Mrs. Alexander* [McNight] *Cameron of Kansas City, MO.,*

> *who are spending a vacation in New York. Mrs. Cameron, the former Viola Gentry, one of the first women pilots at Floyd Bennett Field, has been engaged at Aircraft Accessories Corp., Kansas City, for the past 18 months on precision work. She expects to return to Brooklyn after the war. Viola Gentry organized the Women's International Association of Aeronautics, New York Branch, in 1931 and is its vice president. The dinner in her honor was given by the W.I.A.A.*

After their vacation, Gentry and Cameron returned to Kansas City and continued their jobs at Aircraft Accessories Corporation. Apart from a letter Gentry received from the Department of Commerce on July 1, 1944, which is in the files of the Federal Aviation Administration and documents the issuance of a new pilot's license in the name of Viola Gentry Cameron, there is no other information on her wartime activities. In 1945, Gentry was again absent from the public record except for her and Cameron's entry in the *Kansas City Directory*.

When World War II ended, many wartime factories closed or restructured their workforce. Aircraft Accessories Corporation, however, stayed in business until late 1947 or early 1948, and there is no evidence Gentry or Cameron lost their jobs. Although not documented, it is assumed Gentry and Cameron continued to live and work in Missouri for the remainder of 1945 and throughout the following year.

In the summer of 1946, Thelma G. Hayes (Gentry's sister) and Louise Hayes Campbell (Gentry's niece) visited Kansas City. While there, Thelma received word that her husband, Oakley Cabell Hayes, had suffered a fatal heart attack. The *Martinsville Daily Bulletin* published the sad news on July 8, 1946, and noted that "members of the family are en route to the city." Whether Gentry was in that group is unknown.

In early 1947, Gentry returned to New York and once again became an active participant in the WIAA and attended events hosted by the Ninety-Nines. The first documented function Gentry attended that year was a two-day, all-woman air show in Tampa organized by the Florida Chapter of the Ninety-Nines. Held in March, the purpose of the air show was to raise money for the Amelia Earhart Scholarship Fund. The March 12, 1947 issue of the *Miami Daily News* informed readers of the event. Connie Gee, in her article "Mother of Earhart to Attend Show," reported, "Amy [Otis] Earhart, mother of the world's most famous woman flier, will be met in Tampa...by Mrs. D.C. McRae of Miami, chairman of the Amelia Earhart scholarship fund...Mrs. Earhart will arrive as a guest of

Photograph identified as the Ninety-Nines in Honolulu, Hawaii, 1949. Viola Gentry is seen third from the left. *Courtesy of the International Women's Air & Space Museum, Cleveland, Ohio.*

Eastern Airlines, accompanied by her late daughter's close friend, Viola Gentry Cameron."

As evidenced in photographs, Gentry also attended an event with the Ninety-Nines in 1949. During that year, they traveled from Oakland, California, to Honolulu, Hawaii. In 1935, Amelia Earhart was the first pilot to fly solo from Honolulu to Oakland, and the trip might have been made in remembrance and/or recognition of that flight.

It is unknown how Cameron felt about Gentry's aviation-related activities, which often involved travel between Florida and New York. Her undertakings might have put a strain on their marriage. There is no doubt Gentry and Cameron had marital issues, although the circumstances behind their problems are unknown. Perhaps they had differing opinions on aviation or life in New York. Maybe Cameron was simply unable to fill the void left in Gentry's heart after John R. "Jack" Warren died. No matter the reason, by early to mid-1949, the couple had gone their separate ways, a fact evidenced in the 1949 *Manhattan Telephone Directory*, which documents Viola Gentry Cameron but does not record Alexander.

Samuel and Maydie Gentry, circa 1945. *Courtesy of Helen H. Codling.*

In addition to the emotional strain created by separation and divorce, Gentry also endured the loss of her father. On June 11, 1949, after battling an illness for more than two years, Samuel Garrett Gentry died. Funeral services were held in Danville, Virginia, and he was interred in Highland Burial Park.

Throughout the 1940s, Gentry must have felt as if she had boarded a never-ending roller coaster. She had her ups: a meaningful job, a newfound love and the country's victory in war. Yet with every rise, she had a fall: the loss of a brother-in-law, a troubled marriage and the death of her father.

When Gentry entered 1940, she was a woman who held celebrity status, a woman whose activities were followed and documented by the press. Beginning in 1941 and continuing until the end of the decade, Gentry's name was seen less often; in fact, it was rarely mentioned in the news. Nevertheless, Gentry had not stopped living—or dreaming.

7

Flying Low

1950–59

Viola Gentry, Woman Flier, Back in News
—The Bee, *June 23, 1954*

Newspapers throughout the 1950s proved to be as void of Gentry's name as they had been in the previous decade. However, Federal Aviation Administration records, personal correspondence and oral histories provide a sufficient account of her life during this time period.

Gentry's first order of business for the new year was to finalize her divorce from Alexander McNight Cameron, which was filed in Dade County, Florida, in 1950. Following the divorce, she completed the appropriate paperwork to return her legal name to Viola Gentry. She then notified the Airman's Record Branch, Airman's Division, in Washington, D.C., so the change could be made on her pilot's license.

The second order of business was her occupation. In the early 1950s, Gentry established a pattern of employment that she maintained for more than twenty years. From January until April, Gentry worked as an assistant housekeeper at the Belleview-Biltmore Hotel in Belleair, Florida. For the remainder of the year, she labored at hotels in Florida, New York and other northern states.

Gentry's work schedule did not seem to interfere with her aviation interests, as she was generally in the right city at the precise time to attend meetings and/or events.

In December 1952, Gentry was documented as living in New York. Ruth G. Davis, society editor for the *Brooklyn Eagle*, noted in her December 8

A photograph of the Belleview-Biltmore Hotel, Belleair, Florida, taken between 1979 and 1987. *Courtesy of the State Archives of Florida, Florida Memory.*

"Contemporary Comment" column that "the [WIAA]…honored the famed aviatrix, Miss Viola Gentry at the Wings Club, Hotel Biltmore. Miss Gentry, who organized the club in 1931 at Floyd Bennett Field, returned last week from Florida."

Gentry was indeed a "famed aviatrix." Although she had not partaken of a newsworthy flight in years, she was still remembered for setting the women's solo endurance flight record in 1928. Gentry thought often of her endurance flight, the people who had encouraged her and the contributions each had made to the history of aviation. In an effort to preserve that history, Gentry began collecting aviation-related stories from the friends she loved and respected.

In an effort to document her own aviation record, Gentry addressed a letter to the National Aeronautic Association of the United States (NAA) in Washington, D.C., on April 9, 1953. In it, she asked for a copy of the file that stated she had set the first women's solo endurance flight record. C.S. Logsdon, assistant secretary-treasurer for the NAA, responded on April 10 that they were "unable to locate any references in our files to the solo record flight to which you refer." However, Logsdon told Gentry if she could provide "additional information on the exact record established (distance, speed, etc.) and the date such was accomplished," he would be "pleased to make a further search of our files."

Gentry must have been somewhat annoyed, as her name should have been immediately recognized and the file easily found. Nevertheless, Gentry mailed another letter on May 15, which listed the date of her flight and included copies of the barograph and recording barogram that were carried on the airplane. Nearly a month passed before Gentry received a response. On June 12, Logsdon thanked Gentry for sending the information relating to her flight and said, "For your information many hours research in old NAA files relating to this wonderful achievement included the following…That your solo endurance flight of 8 hours, 6 minutes, 37 seconds was the first official National (U.S.) solo endurance record flight for women…In closing, it is safe to say that you were the holder of the first official (U.S.) feminine solo endurance record, but this performance could not be recognized by FAI at that time, because no international duration classification for feminine pilots existed at that time."

Perhaps after her correspondence with the NAA, the letters of which are in the collection of the International Women's Air & Space Museum, Gentry felt more compelled to write about her flying endeavors, as well as those of other pilots. Yes, there were people who remembered her record, but there were also those who had never heard of it. And for those who did remember, how long would they remember? Would her record eventually be forgotten? How long would the press and the general public recall the triumphs and tragedies of those who took to the sky in the early days of flight? Gentry might have believed that her stories, as well as others, would be disregarded in time.

Determined to preserve the tales of early fliers, Gentry began writing and compiling information. Gentry did not abandon her various hotel jobs in order to write; instead, she worked on the project when time and energy allowed.

In July 1953, Gentry was employed at the Sea Spray Inn, which was located in the town of East Hampton, Long Island, New York. The inn was owned and operated by Arnold Blakeman Bayley, a pilot and aviation enthusiast. Ruth G. Davis, in her July 14, 1953 "Contemporary Comment" column, reported that Gentry was a "receptionist at the Inn."

In the winter of 1953, Gentry returned to Florida and remained there until June 1954. That month, she returned to New York, where she was presented the Lady Hay Drummond-Hay Memorial Trophy. The *Bee* shared news of Gentry's honor, although it incorrectly proclaimed her "Danville-born."

The article "Viola Gentry, Woman Flier, Back in News," which was published on June 23, 1954, stated:

A mid-twentieth-century postcard of the Sea Spray Inn, East Hampton, Long Island, New York. *Courtesy of the author.*

> *Viola Gentry was back in the news today—for the first time in many years.*
>
> *The Danville-born aviatrix, who won international fame as one of the first women fliers, was yesterday presented with the Lady Hay Drummond-Hay silver aviation trophy in New York in recognition of her effort on behalf of women in aviation.*
>
> *The presentation was made for the New York branch of the Women's International Aeronautical Association* [sic] *at the Officer's Club at Floyd Bennett Field.*
>
> *Miss Gentry, is in poor health and her actual presence was not expected. She appeared, however, unexpectedly by air to receive the award…*
>
> *As a pioneer woman flier she earned many trophies and citations and became a well-known figure in American aviation.*

The prestigious and highly coveted trophy was given in memory of Lady Grace Marguerite Hay Drummond-Hay, who had held the position of International WIAA president for ten years. Drummond-Hay, who had also been a personal friend of Gentry's, died in 1946. Because of their friendship, Gentry must have viewed the trophy not just as an award—a material recognition for her achievements in the field of flight—but also as a sentimental treasure.

The Lady Hay Drummond-Hay Memorial Trophy was presented to Viola Gentry by the Women's International Association of Aeronautics, New York Branch, on June 22, 1954, "for her courage and devotion to aviation." *Courtesy of Helen H. Codling.*

Viola Gentry (right) and her niece, Helen Hayes Codling, with the Lady Hay Drummond-Hay Memorial Trophy, 1954. *Courtesy of Helen H. Codling.*

The cause of Gentry's "poor health" was not divulged; however, unless she was paralyzed or in a convalescent state, there was no reason for anyone to assume she would not be present at the event. After all, Gentry had attended the first meeting of the Ninety-Nines with her right arm in a sling and in the company of a nurse.

Gentry was absent from the public record in 1955, but in January 1956, she was back on her job at the Belleview-Biltmore Hotel. That month, she sent a letter, which is in the collection of the Library of Congress, to Elmo Pickerill requesting information regarding his flight from Curtiss Field to Dayton, Ohio; the flight in which she spoke to him through the use of radio equipment. Gentry told Pickerill, "I was sorry not to see you while I was in New York in November but I had so many things to check on—the days just were not long enough."

Apparently, Gentry had spent much of 1955 researching stories for her writing project.

On January 31, 1956, Pickerill responded with a two-page letter, which is also in the collection of the Library of Congress. In addition to the requested information, Pickerill also reminisced about their days at Roosevelt Field. He said, "Enclosed are two pictures of you which I snapped at Roosevelt Field on February 7, 1931. You certainly were one of Roosevelt Field's glamor gals in those days. Nobody ever had more rooters to get well than you had after your serious crash. None of that gang will ever forget you."

When Gentry wrote to Pickerill in 1956, she confirmed her position of assistant housekeeper at the Belleview-Biltmore Hotel—yet she had other duties as well. Gloria Conti Griffin vacationed at the Belleview-Biltmore Hotel with her parents from approximately 1954 to 1962 and, during that time, became acquainted with Gentry. According to Griffin, in a written interview, Gentry often took care of her and her brother while their parents played golf or attended social events at the hotel. Griffin said she recalled "eating lunch with [Gentry] at the pool snack bar" and "her pushing me on a swing."

Whether Gentry was assigned such duties by hotel management or contracted with guests on her own is uncertain. It is also a mystery as to the number of people Gentry interacted with, at that hotel and others, who never knew they were in the presence of such an accomplished woman—a renowned pilot who befriended, flew with, entertained and learned alongside of the legends of aviation.

In November 1959, Gentry again wrote to Pickerill, this time from the Hotel Webster in New York City. She thanked him for a "lovely day" and remarked she "enjoyed her lunch ever so much." The cordial letter, which is in the collection of the Library of Congress, was sent after their attendance of an aviation-related meeting in New York. At the end of her note, Gentry stated, "Was so nice to see all at the meeting—we got home about 12:15, nice ride…did a good job of hanger [sic] flying."

Gentry's use of the term "hangar flying" is noteworthy, as it would become the title of the manuscript she had so diligently begun to write. Her knowledge of and friendship with early pilots would prove to be one of her greatest assets. It would allow her to obtain a professional aviation-related position and give her access to the information she needed to complete her book. Throughout the 1940s and '50s, Gentry was an infrequent blip on the media's radar screen, but in the coming decade, she would light it up and garner attention once again.

8
Flying High
1960–69

Any thing worth while [sic] *is a large undertaking, in all walks of life and even each phase of it.*
—Viola Gentry in a letter to Elmo N. Pickerill, January 24, 1960

Gentry stepped into the new year with both feet on the path of routine. It was a course she had followed for ten years. As the year progressed, however, Gentry's sights focused on a new trail, an upward route that led to the realization of an unfulfilled dream.

Throughout January, Gentry worked at the Belleview-Biltmore Hotel and continued her correspondence with Elmo N. Pickerill. Their conversation, as it always had, regarded aviation and the planning of events relating to that subject. In a January 18, 1960 letter, which is in the collection of the Library of Congress, Pickerill wrote, "Let us know when you come back to New York as we would like to see more of you. You know you have a lot of friends here and they are always glad to see you. You don't seem to belong away down in Florida anyway."

At the completion of her contract with the Belleview-Biltmore Hotel, which ended each April, Gentry indeed returned to New York. It is unknown which hotel Gentry worked at in New York upon her return. She corresponded with Pickerill between the months of June and October on Hotel Webster stationery, so it is possible she was employed there.

When Gentry worked at the Sea Spray Inn in 1953, she established a friendship with the owner, Arnold B. Bayley. In the early part of 1960,

Gentry saw Bayley at a meeting of the Long Island Early Fliers Club at the Westhampton Air Force Base in New York. As reported by Doris Herzig, in the undocumented newspaper article "LIer Roots Double in 'Powder Puff Derby,'" which is in the collection of the Library of Congress, Gentry asked Bayley if she could use his airplane in an upcoming All-Woman Transcontinental Air Race (AWTAR)—commonly referred to as a "Powder Puff Derby."

Bayley, who owned a 260-horsepower Ryan Navion B airplane named *Sea Spray Inn*, told Herzig, "She twisted my arm real hard, and I said 'ouch' and 'yes.'"

Gentry had dreamed of flying in a transcontinental air race since 1929, the year the first Women's Air Derby took place. At the time of that race, however, Gentry was confined to a hospital bed and could only listen to reports of the history-making event.

Thirty-one years later, at the age of sixty-six, Gentry attained her dream through the assistance of Arnold Bayley and Myrtle "Kay" Thompson Cagle. Like Gentry, Cagle was a native North Carolinian. She had learned to fly at the age of twelve and later received her private and commercial pilot licenses. During World War II, Cagle was accepted into the Women Airforce Service Pilots (WASP) program. After the war, she operated Myrtle Airport in Selma, North Carolina; became a member of the Ninety-Nines; and wrote a weekly air column. The column, "Tar Heel Air Currents," was first written for the *Johnstonian Sun* and later for the *News & Observer*.

In her April 24, 1960 column, which appeared in the *News & Observer*, Cagle wrote of her and Gentry's upcoming flight. She said:

> *Viola Gentry and I will race together in the Powder Puff Derby in July. I have always wanted to and so has she. Her "always" goes back farther than mine. It has been Viola's ambition to race since 1929! She was entered in the race of '29 but spent race time in a hospital bed…*
>
> *Viola Gentry is steady and smooth, can handle herself well in the air. She will be a wonderful teammate…to take the controls, run the check list, make navigational notations, and win the race. With her background and experience, and my training, we should finish in the money. It's Captain and Pilot…not the old-fashioned pilot and co-pilot…we're a team.*

On May 24, 1960, the Silver Wings Fraternity, an organization whose members made their solo flights prior to 1933, announced Gentry's participation in the race. Its press release stated:

Left to right: Myrtle "Kay" Thompson Cagle and Viola Gentry prepare for the 1960 All-Woman Transcontinental Air Race. *Courtesy of the International Women's Air & Space Museum, Cleveland, Ohio.*

A Silver Wing, Viola Gentry, will fly #51 in the All-Woman Transcontinental Air Race from Torrance, California to Wilmington, Delaware, July 9–13, 1960. The plane which she will use is a Navion, lent her by Arnold Bayley, of Sea Spray Inn, East Hampton, N.Y.

Left to right: Myrtle "Kay" Thompson Cagle, Viola Gentry, Arnold B. Bayley, Shirley Marshall and Jane Griese. Bayley loaned Cagle and Gentry his airplane for the 1960 All-Woman Transcontinental Air Race and sponsored the team of Marshall and Griese by financing their flight suits. *Courtesy of the International Women's Air & Space Museum, Cleveland, Ohio.*

> *Her flying partner is Myrtle Cagle—another Tar Heel. Myrtle is an aircraft instructor and flies from Myrtle Airport, Selma, North Carolina.*
>
> *Viola is known by all Silver Wings as the gal who learned to fly at Curtiss Field, Long Island, New York. She established the first official solo record for women on December 20, 1928. On July 28, 1929, flying with Jack Ashcraft in an attempt to establish a new Refueling Endurance Flight Record, they "cracked-up." Jack was killed and Viola spent 22 months in hospitals. But Viola flew again—and all Silver Wings join in cheering her on to win.*

In a letter to Elmo N. Pickerill, dated June 25, 1960, Gentry shared news of, and her excitement over, the race. The note, which is in the collection of the Library of Congress, read:

Dear Pick,

...Leaving East Hampton June 28 for California and the Powder Puff Derby—landing at Wilmington, Del. July 13—then back to Sea Spray for a rest.

Thrilled over the race, Mr. Bayley is lending us his Navion—we have a good chance to win as we have a landing handicap speed of 150 pmh. [sic]

Will see you at the meeting...
Viola Gentry

Cagle and Gentry took flight from Torrance Municipal Airport in Torrance, California, on Saturday, July 9. From there, they flew approximately 2,700 miles to Wilmington, Delaware, in an attempt to win the 1960 AWTAR. At the completion of the race, seventy-nine airplanes entered and seventy-six finished, scores (the handicap speed of an airplane was subtracted from its average ground speed) were tabulated and a winner declared. Cagle and Gentry did not win the race but placed forty-ninth.

Following the awards banquet, which was held on Friday, July 15, in Wilmington, Delaware, Gentry and Cagle traveled to the Sea Spray Inn. There, the women, along with Arnold Bayley and other pilots, relaxed and enjoyed each other's company. On Sunday, the two women partook of a "Powder Puff Dinner" that included selections such as "Gentry fruit cup supreme," "Myrtle and Walt [Myrtle's husband] Cagle pot au feu," "coconut powder puff layer cake" and "Navion sundae."

The following month, Gentry was honored for her participation in the air race by the Long Island Early Fliers Club. Elmo N. Pickerill, president of the club, mailed postcards, one of which is in the collection of the Library of Congress, to members. It read:

Sunday, August 14, 1960 has been designated
Viola Gentry Day
by the Long Island Early Fliers Club to celebrate her
participation with Mrs. Myrtle Cagle in the recent
POWDER PUFF DERBY from California to Atlantic coast.
Meeting will start promptly at 2:00 p.m. and dinner at
5:00 p.m. in the Officers Club, Suffolk Air Force Base,
Westhampton, N.Y. Come early and bring a friend.
Special permission granted by the Commanding Officer
for members of the Ninety-Nines, OX5 Clubs, and the
Antique Airplane Ass'n to land their privately owned

> *airplanes on the Air Force Base, starting around noon, all of whom are invited to attend…*
>
> *Happy landings.*
> *Elmo N. Pickerill*
> *President*

For the remainder of the year, Gentry continued to work at various hotels and participate in meetings and/or events of the aviation organizations to which she belonged. She also planned and made arrangements to fly in the 1961 AWTAR.

During the 1960 AWTAR, Arnold Bayley sponsored not only Cagle and Gentry in the race but also the team of Shirley Marshall and Jane Griese. Bayley had met Marshall and Griese in Tucson, Arizona, in the fall of 1959 and agreed to finance their flight uniforms.

Marshall and Griese stayed at the Sea Spray Inn the weekend after the AWTAR, just as Gentry and Cagle had. This is evidenced by their autographs on Gentry's "Powder Puff Derby" menu. Although Gentry and Marshall likely knew each other prior to the AWTAR (both were members of the Ninety-Nines), they might have become better acquainted during their stay at the inn. There is no doubt they remained in touch with each other, as on May 20, 1961, the *Tucson Daily Citizen* announced:

> *Tucsonian to Enter Air Race*
>
> *Mrs. Shirley R. Marshall…will be flying in her third Powder Puff Derby, the all-woman transcontinental air race slated July 8 through 12.*
>
> *Copilot for Mrs. Marshall will be Miss Viola Gentry, a member of the New York Women's Press Club and a charter member of the Ninety-Nines.*
>
> *The Marshall-Gentry team…will be flying a Piper PA-22 with a 150 horsepower rating.*
>
> *The race will begin at noon (EDT) Saturday, July 8, in San Diego, Calif., and will end at noon (EDT) Wednesday, July 12, in Atlantic City, N.J.*

Gentry had waited thirty-one years to participate in an AWTAR, but she only had to wait a year to compete in her second one. As before, parties were held to celebrate Gentry's entry into the race, and the press printed articles regarding the "pioneer flier."

In June, Shirley Marshall, along with her husband Art, gave a grand party in Gentry's honor. The invitation, which is in the collection of the Library of Congress, read:

You are invited to a punch party
The guest of honor will be Viola Gentry
Member OX-5 Club
Charter Member 99's
Member W.I.A.A.
Member Long Island Early Fliers Club
and
Co-pilot in "Desert Honey" for the 1961 AWTAR
Saturday, June 24, 1961
5–8 p.m.

Desert Honey was the name of Shirley Marshall's Piper PA-22 Tri-Pacer airplane and the one she flew in the 1960 AWTAR. As before, Arnold Bayley sponsored Marshall and Gentry in the race, only this time he supported them as a team, not as individuals. The Tucson Airport Authority also assisted the duo by providing their fuel. The AWTAR began on Saturday, July 8, and was slated to end on Wednesday, July 12. Bad weather, however, forced officials to postpone the completion of the race for forty-eight hours. As a result, the AWTAR officially ended on Friday, July 14. Out of ninety-seven entrants, seventy-seven—including Marshall and Gentry, who placed forty-sixth—completed the approximate 2,700-mile air race.

Prior to the race, on June 26, 1961, the *Tucson Daily Citizen* published an article on Gentry. The story "Life Began with No. 1822 For Daring Woman Pilot," by Renee Chipman, told of Gentry's early flying escapades, as well as her struggles to finance them. In the article, Gentry proclaimed, "I was born in 1822!"

Never before had Gentry so well expressed, in words, her passion for flying. Chipman said the remark was "Viola's way of saying that age is not measured by the chronological passage of years, but that she did not start to live until she had qualified for her private license No. 1822."

At the end of the write-up, Chipman asked Gentry to name her greatest flying moment. Gentry responded, "My greatest moment? Without a doubt my first solo…I can't put into words what I felt. It was not so much a sense of accomplishment—there was something else, as if I were up there—alone—with Him."

In addition to providing profound physical and spiritual joy, flying also afforded Gentry the comradery of many good friends. Several years after competing in the 1961 AWTAR, Gentry composed a short biography, which is in the collection of the Ninety-Nines, Incorporated,

Left to right: Viola Gentry and Shirley Marshall stand in front of Marshall's Piper PA-22 Tri-Pacer airplane prior to the 1961 All-Woman Transcontinental Air Race. *Courtesy of the International Women's Air & Space Museum, Cleveland, Ohio.*

that summed up those feelings. She said, "If I never fly again, [I]…had enough fun that year [1961] to last my next 50 years. God has blessed me. It has been a privilege to have known—and still know—many of the great pilots of yesteryear and today."

Gentry did not compete in the 1962 or 1963 AWTAR. She had planned to fly in the 1963 AWTAR but was unable to secure a sponsorship. Nevertheless, in 1963, Gentry teamed up with Shirley Marshall again. This time they would not fly in an air race but in celebration of Amelia Earhart.

On July 24, 1963, in observance of Amelia Earhart's birthday, the United States Post Office issued an eight-cent domestic airmail stamp. The issuance of the stamp was celebrated in Atchison, Kansas, Earhart's hometown. According to the July 21, 1963 *Anderson Herald* article "Nation Joins in Earhart Birthday Fete," Gentry, along with Shirley Marshall and Pat Nolen, flew to Atchison with a "replica of a bronze Amelia Earhart plaque" that "hangs in the National Air Museum of the Smithsonian in Washington." Once in Atchison, Gentry presented the plaque to Purdue University. Following the presentation and other events, Gentry and six other charter members of the Ninety-Nines left Kansas to deliver the First Day Covers to locations across the country.

An undocumented newspaper clipping, which is in the collection of the Library of Congress, records Gentry's participation in the event. The article "Pioneer Aviatrix Viola Gentry Joins Amelia Earhart Salute," by George Carroll, stated:

The United States Post Office issued an Amelia Earhart eight-cent domestic airmail stamp on July 24, 1963, in honor of Earhart's birthday. Seven charter members of the Ninety-Nines, including Viola Gentry, flew the First Day Covers to locations across the country. The covers were sold to benefit the Amelia Earhart Scholarship Fund. *Courtesy of the author.*

Left to right: Shirley Marshall, Olive Ann Beech, Viola Gentry and Pat Nolen with a bronze plaque of Amelia Earhart, July 23, 1963. *Courtesy of the International Women's Air & Space Museum, Cleveland, Ohio.*

> *Pioneer aviatrix Viola Gentry of East Hampton, L.I., takes to the air Wednesday in a flying salute to a famous friend, Amelia Earhart.*
>
> *She's one of seven charter members of the 99s,* [an] *international society of 1,600 licensed pilots who'll radiate off to various cities of the country from Amelia Earhart Airport in Atchison, Kan., the town where Miss Earhart was born.*
>
> *Their planes will be laden with official first day covers bearing the new 8¢ Amelia Earhart domestic airmail stamps. These are being sold at a dollar apiece to enlarge the Amelia Earhart Scholarship Fund.*

In addition to flying around the country, Gentry, Marshall and Nolen—as reported on July 23, in the *Albuquerque Tribune* story "They Remember Amelia Earhart"—flew "aboard Pan American and Australia airlines to Australia, New Guinea, Hawaii, Korea, Puerto Rico and South America, to distribute the commemorative stamps and cachet covers."

Gentry's participation in the Earhart festivities must have reminded her of the obstacles women faced when they first took to the sky and of their achievements

Left to right: Captain Powell of Pan American World Airways, Viola Gentry, Shirley Marshall and Pat Nolen. Gentry is seen presenting Captain Powell with an Amelia Earhart First Day Cover, July 1963. *Courtesy of the International Women's Air & Space Museum, Cleveland, Ohio.*

and contributions to the field of aviation. She likely thought of Earhart's solo flight across the Atlantic Ocean and how it had inspired young women across the globe. Perhaps it made her think of other American women—pilots who desired to fly higher than any woman before them but were not allowed to do so. Whatever thoughts might have been stirred by the Earhart event, it is clear Gentry was not happy with the National Aeronautics and Space Administration's (NASA) stance on female astronauts. The article "Woman Flier Hits NASA on Policies," published in the *Pampa Daily News* on November 24, 1963, shared Gentry's thoughts on the matter. It read:

> *Dallas (UPI)—Viola Gentry renewed an old squabble Thursday and took a swipe at the federal space agency for not having as much faith in American women as they do in monkeys.*
>
> *Miss Gentry, who flew Curtiss bi-planes before most astronauts were born, told* Dallas Times Herald *science editor Bob Fenley it burns her up because the United States will not let women be astronauts…*

> *Now desk clerk of a resort hotel at East Hampton, L.I., Miss Gentry would only say that she is "past 60."*
>
> *But she still flies, in spite of her apparent non-flying job.*

Gentry's opinion did not change NASA's policies. Nonetheless, she had voiced her thoughts on the matter.

For the remainder of 1963 and throughout 1964, Gentry focused once more on acquiring stories from early pilots. She had become a collector of aviation memories. Gentry not only gathered stories but also disseminated facts, and items of her own, to institutions and organizations where they were sure to be preserved.

On October 24, 1963, the *Warrensburg–Lake George News* announced that Gentry had presented a special book to the Richards Library of Warrensburg, New York. The article "Rare Floyd Bennett Book Given Library" said:

> *A rare volume, entitled "Floyd Bennett," and written by Mrs. Cora Bennett, widow of the famed flyer whom Warrensburg claims as its most distinguished son…was the gift of Miss Viola Gentry…a lifelong friend of Mrs. Bennett, and an enthusiastic student of the history of aviation.*
>
> *Miss Gentry came to Warrensburg several weeks ago to discuss with the Floyd Bennett Memorial Committee a project to make a bronze plaque of Bennett. While here she learned that the library had been vainly seeking a copy of Mrs. Bennett's biography of her husband.*
>
> *"I have a copy which I shall be glad to donate to the library," Miss Gentry said.*
>
> *"I could hardly believe our good fortune," Mrs. Cameron said. "I had just about given up hope that we would ever have, in Floyd Bennett's hometown, a copy of this rare volume. But thanks to the generosity of Miss Gentry we now own one. We shall never forget her kindness."*

It is interesting to note that Cora Bennett presented her book *Floyd Bennett* to President Herbert Hoover in 1932. Gentry accompanied Bennett on her flight to Washington, D.C., and stood beside President Hoover as he received the publication.

The early 1960s had proved to be an exhilarating time for Gentry. She had spent a considerable amount of time in the air, fulfilled her dream of competing in an AWTAR and recaptured the media's interest. The remaining years of the decade produced a similar outcome for Gentry, as she fulfilled another dream and was recognized by several organizations for her contributions to aviation.

Viola Gentry's OX5 Aviation Pioneers membership card. *Courtesy of Helen H. Codling.*

Left to right: Betty Robertson Uhl, Viola Gentry and Tiny Broadwick. Gentry was presented the Tiny Broadwick Award by the OX5 Club of America in October 1965. Tiny Broadwick, a native North Carolinian, was the first woman to parachute from an airplane. *Courtesy of the International Women's Air & Space Museum, Cleveland, Ohio.*

Ed Harvey, associate manager of the Sea Spray Inn, congratulates Gentry for receiving the Tiny Broadwick Award. Photograph likely taken at the Sea Spray Inn in 1965. *Courtesy of the International Women's Air & Space Museum, Cleveland, Ohio.*

In October 1965, a national reunion of the OX5 Aviation Pioneers was held in Chicago, Illinois, by the OX5 Club of America. Founded in 1955, the OX5 Club of America celebrates the Curtiss-built Model OX5 V8 engine and the pilots who flew them. At the reunion, Gentry was presented with the prestigious Tiny Broadwick Award, a recognition she well deserved. The words on the plaque read:

TINY BROADWICK AWARD
PRESENTED TO
VIOLA GENTRY
HER ACTIVITIES IN THE PROMOTION AND DEVELOPMENT
DEVELOPMENT
OF AVIATION HAVE ENCOURAGED WOMEN
THROUGHOUT THE WORLD TO PARTICIPATE IN THE
FIELD OF AERONAUTICS
OX5 CLUB OF AMERICA
1965

The following month, Gentry, along with Felicity Buranelli and Arnold Bayley, traveled to Austin, Texas, where they attended—as reported in the WIAA newsletter of January 1966, which is in the collection of the Ninety-Nines, Incorporated—"the dedication of the Aviation Library at the University of Texas by George Haddaway, publisher of *Flight Magazine*."

While there, "a ten-inch bronze plaque in memory of Vincent Burnelli was unveiled by Arnold Bayley, pilot, who represents many flying clubs and who was donor of the plaque. Later they went to Temple, Tex. where Viola Gentry, donor, unveiled the Vincent Burnelli sister plaque in his home town."

Gentry began working with Felicity Buranelli, who was the creator of the Medal-of-the-Month Club in 1960. Buranelli started producing medals in 1941 of select aviation pioneers. Her idea was to market the medals to children, who she hoped would be inspired to learn more about American history. The first medal Buranelli commissioned was of Wilbur and Orville Wright. The original was a ten-inch bronze plaque, from which

Left to right: Viola Gentry, Felicity Buranelli and Colonel Harold Woodruff, 1962. Buranelli holds a bronze plaque of Lieutenant Colonel John H. Glenn Jr. *Courtesy of the International Women's Air & Space Museum, Cleveland, Ohio.*

one-and-a-quarter-inch bronze medals were struck. The process, however, was slow going, and in 1962, she had produced only eight medals. They included: Wilbur and Orville Wright, Captain Edwin C. Musick, Amelia Earhart, Lieutenant Commander Frank Hawks, Brigadier General William Lendrum Mitchell, Glenn H. Curtiss, Brigadier General Frank P. Lahm and Lieutenant Colonel John H. Glenn Jr. At the completion of her aviation series, Buranelli created two additional medals: one in honor of her brother, Vincent Burnelli (Vincent spelled his last name without the letter *a*), and the other for Bernard R. Hubbard, an explorer, photographer, lecturer and Jesuit priest.

Gretchen S. Black, a member of Chapter Goldsboro Delta Lambda, North Carolina and International Beta Sigma Phi, wrote a short biography of Gentry on January 25, 1966, titled "For Consideration as an International Honorary Member of Beta Sigma Phi."

The biography, which is in the collection of the Ninety-Nines, Incorporated, provides information regarding Gentry's association with Buranelli. In it, Black wrote:

> *In 1960* [Gentry] *started working for Felicity Buranelli, who* [sic] *famous brother Vincent, designed, built, and flew the* [Burnelli] *airplane. Miss Buranelli is originator of the bronze medals of famous flyers and it is Viola's job to travel over the country, to sell this idea to the States, from where many of these famous old pilots came from. The small medals are about the size of a dollar and are sold as souveniers, but the huge bronze plaques, are installed in a public building, such as that of Capt. Edwin C. Musick, whose plaque hangs in a place of honor at Pan American Airways Terminal at Kennedy Field in New York. Viola has a collection of these medals, which she has had made into a necklace and you seldom see her without it.*

Whether Gentry received a commission on the Buranelli medals she sold or took orders for is unknown. It is certain, however, that her connection with Buranelli provided opportunities to revisit old friends and spend time at various aviation-related institutions. This, in turn, likely aided Gentry in her search for information on early pilots.

In regard to Gretchen Black's request for Gentry to become an International Honorary Member of Beta Sigma Phi, her appeal was approved. Black said of Gentry, "She is a real Beta Sigma Phi, if I ever knew one, and surely epitomizes our motto of Life, Learning and Friendship."

Viola Gentry frequently wore a necklace made from the aviation medals produced by Felicity Buranelli. This photograph shows Gentry wearing the necklace in 1971. *Courtesy of the International Women's Air & Space Museum, Cleveland, Ohio.*

Gentry traveled to Goldsboro, North Carolina, in April to accept the honor. Elga Nichols, in her April 28, 1966 *Goldsboro News-Argus* column "Social and Personals," stated:

> *Miss Gentry to Receive Two Rituals*
>
> *Miss Viola Gentry, veteran woman pilot and a charter member of the first 99's Chapter, which is an organization of women pilots, is a guest of Mr. and Mrs. Grant Black on Mulberry Street.*

> *On Friday night, Miss Gentry will receive both the Pledge Ritual and the Ritual of Jewels at the home of Mrs. Black. These rituals are given to qualify the recipient to receive the distinctive honor of being made a Beta Sigma Phi International Honorary Member, which will be given her Saturday night at the Founder's Day Banquet at Del Monty's in the Pioneer Room.*
>
> *Mrs. John B. Chase, sponsor for Delta Lambda Chapter, is also an honorary member, along with Eleanor Roosevelt, Margaret Chase Smith, Mrs. Calvin Coolidge and many other notable women. Both Mrs. Chase and Miss Gentry's name will be listed in the Beta Sigma Phi WHO's WHO directory.*

Gentry no doubt beamed over her accolades, yet an opportunity presented to her in the spring of 1967 likely surpassed them all. An undated clipping from the *Austin American* newspaper titled "Noted Aviatrix Named Consultant to UT Collection," which is in the collection of the International Women's Air & Space Museum, announced:

> *A lively little lady who has been an aviatrix since 1925, when she made her first solo flight and her family "wrote…off as crazy," has been appointed by the Humanities Research Center as a consultant to the History of Aviation Collection at the University of Texas.*
>
> *Miss Viola Gentry of Miami, Fla., will take up her duties for the university immediately, traveling throughout the United States in search of materials from veteran pilots to add to UT's already extensive collection.*
>
> *The new consultant has wide acquaintances with many of the "Early Birds," fliers who earned the name by having soloed before 1916. Many of them will be seeing Miss Gentry in her official capacity in the near future. In her visits with the pilots, she will be looking for correspondence, pictures, scrap books, log books, flying records and other items.*

Another undocumented clipping from the same collection, titled "Long-Time Pilot Becomes Curator," reported:

> [Gentry's] *lasting zeal for aviation is responsible for a turn of events in her life—she has been named the curator of the history of aviation collection at the University of Texas…*
>
> *Miss Gentry's first assignment as curator is to travel over the country contacting Early Birds…*

This photograph accompanied the press release that announced Gentry's appointment as consultant to the History of Aviation Collection (HAC) at the University of Texas at Austin in May 1967. Gentry is seen standing next to George Haddaway, who established the HAC in 1963. *Courtesy of the International Women's Air & Space Museum, Cleveland, Ohio.*

> *Today, the small, gray-haired woman talks intently for the cause of aviation. She is delighted with her new job and with the University of Texas campus which she finds "beautiful."*

At long last, Gentry had acquired a position that dealt solely with aviation, an appointment that tasked her with collecting the historic memories and relics of early pilots. It was a job for which she was well suited—a project whereby her knowledge, experience and friendships would aid in the preservation of aviation history.

Gentry's role as consultant to the History of Aviation Collection (HAC) at the University of Texas at Austin began on May 1, 1967. Immediately she began mailing notices of her appointment to Early Bird friends and acquaintances. Her letters told of the HAC and its desire to "honor and pay tribute to…pioneer pilots."

She then encouraged the recipients of her letter to donate their aviation-related correspondence, pictures, log books, etc. to the HAC or allow her to make copies of them. Gentry also informed Early Birds they could receive a binder from the HAC, if interested, in which to insert their biographical information. Once completed and returned, the binders would be permanently filed in the HAC Library.

In addition to mailing letters to Early Birds, Gentry also traveled the country in search of information. From coast to coast, Gentry visited countless towns in her hunt for surviving Early Birds and their flight relics. Many newspapers reported Gentry's visit, as well as the purpose of her stay, to their readers. On November 2, 1967, the *San Jose Mercury* published the article "Flight Relics Sought Here for 'Early Bird' Library," which told of Gentry's arrival in California, her aviation background and her quest to collect donations for the HAC. In addition, the story also documented Gentry's work on her own publication. It said, "Besides her searching out Early Birds, she is writing a book, "Hangar Flying," revolving around the aviation history of better known US fliers. A friend, Walter Bohrer of Aptos, himself an author of several books on aviation, is checking her script for her."

Despite being in her seventies, Gentry not only traveled, collected and organized material for the HAC but also labored at her book, fulfilled her membership duties with aviation-related organizations and logged enough flight time to maintain her pilot's license. At each of those tasks, Gentry excelled. As a matter of fact, the book *Aeronautics and Space Flight Collections*, which was edited by Catherine D. Scott, spoke of Gentry's success in acquiring material for the HAC. It stated:

> *In 1967, Ms. Viola Gentry…made a tour of the United States soliciting memorabilia from the Early Birds, World War I aces and other aviation notables. This brought a flood of significant contributions from 1967 through 1974, which included historical data and books from the Aero Club of Long Island and records of such Early Birds as Bernetta Adams,*

Left to right: Mrs. Giuseppe Bellanca (Dorothy), George Haldeman and Viola Gentry at the Early Birds Reunion in Miami, Florida, 1967. *Courtesy of the International Women's Air & Space Museum, Cleveland, Ohio.*

> *Thomas Benoist, "Tiny" Broadwick, Giuseppe M. Bellanca, Glenn Curtiss, Robert Fowler, Lyman Gilmore, Beckwith Havens, Ed and Milton Korn, Lestere Miller, Willy Ober, Elmo Pickerill, Cal Rodgers, Blanche Stuart Scott, H. Roy Waite, Clifford L. Webster, Horace B. Wild, and a host of others.*

In addition to collecting material from Early Birds, Gentry also procured books and information from Stella Randolph for the HAC. Randolph was a lawyer who became interested in the flights of Gustave Whitehead. After researching Whitehead, collecting artifacts and interviewing witnesses who had seen his machine take flight in 1901, Randolph became convinced that he, not the Wright brothers, had made the first powered, sustained and controlled flight. As a result, she published several books in an effort to prove it and have him recognized as the first to fly.

Randolph's claims regarding Whitehead were, and still are, controversial. Nevertheless, Gentry understood her desire to see Whitehead acknowledged for his contributions to aviation. In a letter to Randolph dated October 10, 1969, which is in the History of Aviation Collection, Special Collections Department, in the McDermott Library at the University of Texas at Dallas,

Gentry encouraged her with the words, "Carry on…let no one stop you unless he can prove that you are wrong."

Gentry's words would not have been spoken lightly, as she had carried on time and time again; she had never allowed anyone or any circumstance to deter her from attaining her dreams. Yet as Gentry headed toward 1970, she faced a new challenge: an age-related hindrance with the ability to ground her for life. In October 1969, Gentry attended a meeting of the OX5 Club of America in Fort Worth, Texas. While there, a reporter with the *Fort Worth Star-Telegram* interviewed Gentry. The story "Woman Pilot Is Sweating Physical," which was published on October 24, 1969, documented Gentry's failing eyesight. Gentry said, "My eyes don't see so well…But I will continue to fly if I pass that physical next Saturday."

The article also said that "one of her biggest thrills" was being "asked to collect the history of aviation for the University of Texas at Austin."

Gentry passed her physical and glided into 1970 with an intact set of wings. She remained persistent in her search for aviation-related memorabilia and worked without fail on her own projects. The 1960s had proved to be an eventful decade for Gentry, and so, too, would the 1970s and '80s.

9

Final Flight

1970–88

Viola Gentry, one of aviation's selfless souls...
—Charles E. Planck, Women with Wings, *1942*

Gentry's efforts in collecting material for the History of Aviation Collection (HAC) at the University of Texas at Austin continued to enrich its holdings. In a letter dated March 20, 1970, which is in the collection of the International Women's Air & Space Museum, George Haddaway said, "Dear Friend of the Aviation Library: Our annual meeting in Austin January 23–24 was a great success...Viola Gentry, our traveling 'Bird Dog' gave an optimistic report on her travels in 1969 and told of her expectations for 1970. Early Birds and OX-5 members have been especially gracious to her in cooperating with the search for historical materials."

Haddaway's use of the term "Bird Dog" was most appropriate, as Gentry had successfully tracked down Early Birds across the country. It was a hunt she participated in throughout the early 1970s.

While working for the HAC, Gentry continued to write *Hangar Flying*. She also composed several pieces of fiction. Prior to April 1970, Gentry wrote *The Birds' Musical*, a charming tale about sparrows on a Carolina farm, which she printed in the form of a small booklet. The exact date *The Birds' Musical* was printed is unknown; however, in April 1970, a man named George (last name unknown) wrote a letter, which is in the collection of the International Women's Air & Space Museum, to Gentry that said "enjoyed [the] book about the birds."

In addition to *The Birds' Musical*, Gentry composed three other nature-based short stories: *The Baby Cardinal*, the story of a fledgling cardinal who flies into a house and cannot find his way out; *Chirpy and Croakey*, which explores the friendship between a cricket and small frog; and *The Patch*, a tale about a black-eyed pea and string bean.

Gentry enjoyed writing and, in most instances, appreciated being written about. However, in November 1970, Gentry's name became associated with a newly released publication, a book she had neither written nor read, that placed her within a whirlwind of controversy. The book, *Amelia Earhart Lives* by Joe Klaas, documented Major Joe Gervais's ten-year search for the legendary female pilot. Gervais's research began in 1960, when he, along with other U.S. Air Force personnel, began Operation Earhart, the goal of which was to determine what had happened to Amelia Earhart.

Since 1937, Gentry had held to the belief that Amelia Earhart had landed on an unknown island and that she did not crash. She had long hoped her friend would be found alive. She wanted—like so many others—for the mystery of Earhart's whereabouts to be solved. As a result, Gentry supported and remained interested in search efforts.

In August 1965, Gentry invited Gervais to present his findings on Earhart to members of the Long Island Early Fliers Club at the Westhampton Air Force Base Officer's Club. She also asked him to attend a reception at the Sea Spray Inn. Gervais agreed to both invitations.

At the Sea Spray Inn, Gervais enjoyed meeting, talking with and taking photographs of early pilots. According to Klaas in *Amelia Earhart Lives*, Gervais was being introduced to Carl A. "Slim" Hennicke, a pioneer pilot and founder of the Long Island Early Fliers Club, and Elmo Pickerill when:

> *Viola Gentry...glanced into the reception room...*
>
> *"Why there's Irene Bolam," she said...with surprise and awe. "It really is Mrs. Bolam," she repeated.*
>
> *Gervais turned to look at the...woman who had just entered the room. A chill ran through Gervais and he trembled slightly.*

When Gervais saw Bolam, he immediately saw Earhart. To him, there was more than a resemblance between the two. Gervais asked Gentry to introduce him to Bolam, which she did. After conversing with Bolam, Gervais was convinced she was Amelia Earhart. He further deduced Earhart had secretly returned to the United States and assumed the identity of Irene O'Crowley Craigmile Bolam. (Irene's maiden name was O'Crowley. She married Charles Craigmile and, following his death, married Guy Bolam.)

After the meeting of the Long Island Early Fliers Club, Gervais continued his research, focusing specifically on the early life of Bolam, as well as her friendship with Earhart and Gentry. Five years later, his research was documented in the book *Amelia Earhart Lives.*

Upon the book's release, newspapers reported its hopeful, yet sensational, claims. On Tuesday, November 10, 1970, the *Washington Post* published the article "Amelia Earhart Reported Alive," which stated, "Lt. Col. Joe Klaas and Maj. Joseph Gervais said in a news release they believe a woman calling herself Mrs. Guy Bolam, and claiming to have flown with Miss Earhart, actually is the famed aviatrix herself. They base their belief largely on physical resemblance and Mrs. Bolam's secrecy about her past."

Nancy Taylor and Don Bedwell, staff writers for the *Miami Herald*, interviewed Gentry for the article "Amelia May Not Have Been Killed, but Friends Say She's Not Irene." The article, which was published on November 14, 1970, stated:

> *Like the retired Air Force officer who for 10 years researched "Operation Earhart," Miss Viola Gentry of Miami isn't convinced Amelia Earhart really died when her plane disappeared in the Pacific in 1937.*
>
> *But when Maj. Joe Gervais (USAF Ret.) suggests in his book, written by Joe Klaas…that the famed aviatrix may be alive and living in Monroe, N.J., as Irene (Mrs. Guy) Bolam, Miss Gentry says, "Why, that's ridiculous!"…*
>
> *"I've known Irene Bolam since 1930 and I knew Amelia Earhart, too, when we all flew from Roosevelt Field. They certainly are not the same woman," says Miss Gentry…*
>
> *MISS GENTRY, an unofficial historian of aviation…knows researcher Joe Gervais.*
>
> *"I've known his family in Massachusetts a long time…"*
>
> *Viola is mentioned frequently in the book, which she hasn't read, but she says she "had no idea Joe was doing this. If I had, I would have rapped his knuckles."*

Throughout November 1970, newspapers continued to report on the book and provided statements from those who defended and denounced its claims. Bolam publically denied she was Amelia Earhart and eventually sued the author and publisher for defamation.

Viola Gentry, 1971. *Courtesy of Helen H. Codling.*

Gentry likely never imagined that a chance meeting between two individuals would result in such an assertion or that she would be considered knowledgeable of the supposed ruse. Nevertheless, as 1971 progressed, the media turned their attention to other stories, and Gentry set her sights on the 1972 All-Woman Transcontinental Air Race (AWTAR).

At the age of seventy-seven, Gentry continued to fly airplanes in order to maintain an active pilot's license. When she renewed her license in 1972, however, it came with a new limitation. As documented on her May 2, 1972 medical certificate, which is in the files of the Federal Aviation Administration, Gentry was required to "wear glasses for distant vision while flying."

In June and July, Gentry's name, and ear-to-ear grin, graced the pages of several newspapers, as reporters told the story of an elderly woman who, after flying for nearly a half century, still enjoyed parting the clouds and competing in air races. A headline from the July 4, 1972 *Redwood City Tribune* declared, "72 Years Old and Ready to Race/They Can't Keep Viola Gentry Out of the Sky."

A staff writer for the *Miami Herald*, in the article "She's Been Up in the Air—In Planes—for 47 Years," which was published on June 16, 1972, called Gentry "the longtime lady aviator" who was born on "June 13, 1900." In response to the statement, Gentry said, "But that's just when I was born. I'm not that old."

Had the media known Gentry's real age, she might have been assailed by the press before, during and after the race.

The *Miami Herald* article also stated, "[Gentry] has two superlative…jobs. Works every summer at the Grand Hotel in Mackinac, Mich., which she says has the longest porch in the world, and every winter at the Bellevue Biltmore [*sic*] at Clearwater—'the largest wooden structure in the world.'

During the summer of 1972, Gentry apparently took an extended vacation from her duties at the Grand Hotel in order to participate in the

Postcard of the Grand Hotel, Mackinac Island, Michigan, postmarked 1946. *Courtesy of the author.*

Left to right: Viola Gentry and Ruth Johnson, 1972. *Courtesy of the International Women's Air & Space Museum, Cleveland, Ohio.*

AWTAR. By the end of June, Gentry had arrived in California, which was the starting point for the race.

On July 7, Gentry and Ruth Johnson, a former member of the Women Airforce Service Pilots (WASP) who operated a flying school near Los Angeles, took off in San Carlos and headed toward Toms River, New Jersey. They flew a Piper Comanche airplane, which had been marked with the words "Salute Clara Adams and Arnold Bayley."

Bayley, who had been Gentry's employer, friend and race sponsor, passed away in 1970. Adams, the first woman to fly across the Atlantic Ocean (as a passenger) on the Graf Zeppelin, was a dear friend of Gentry's who died in 1971.

Angela Masson, the first woman to be type-rated on the Boeing 747 and the second female pilot hired by American Airlines, also flew in the

Viola Gentry (center) and Angela Masson at the All-Woman Transcontinental Air Race Gala Awards Banquet in Point Pleasant Beach, New Jersey, July 12, 1972. *Copyright Airplanet Productions, Incorporated.*

1972 AWTAR. She said that 105 airplanes entered the race; however, 5 were disqualified, and 3 dropped out. At the completion of the approximate 2,600-mile race, Gentry and Johnson placed twenty-fourth.

Masson, who was flying her first AWTAR, flew solo and set a record for being the youngest person to fly coast to coast in a high-performance aircraft, a Bellanca Super Viking. She placed fiftieth in the race. In regard to Gentry, Masson said, "I didn't have much conversation with Viola…Just regular chatter/small talk when we passed at the gas stops…She was super sweet, spunky, and a great sport."

At the Gala Awards Banquet, which was held at the Beacon Manor Hotel in Point Pleasant Beach, New Jersey, on July 12, 1972, the master of ceremonies introduced Gentry and Masson as "the oldest and youngest participants." Gentry was "seventy-two" and Masson twenty-one.

The following day, the *Ocean County Review* ran an article about the race. The untitled clipping, from the collection of Angela Masson, said:

> *Everybody fell in love with co-pilot Viola Gentry, of Miami, Fla…Viola says she is "42 and holding." You better believe it…she was born* [in]

Postcard of the Monmouth Hotel, Spring Lake, New Jersey, circa 1970s. *Courtesy of the author.*

> *1894! She soloed in 1925 with her first official solo record in 1928. There were no licenses "in those days" and Viola "flew free" until #1822 was issued in her name...Asked if she had heard from her friend Amelia Earhart lately, Viola deadpanned "no, there must be something wrong with the mail." On the ground this amazing woman is assistant housekeeper for resort hotels and was housekeeper for three years at the Monmouth Hotel in Spring Lake, N.J.*

For the first time in nearly fifty years, Gentry's accurate date of birth had been revealed. One has to wonder if Gentry—who might have been overwhelmed by the race, either with excitement or exhaustion—had inadvertently revealed that piece of information to the reporter. One will never know, but in interviews and articles that followed the 1972 AWTAR, Gentry once again gave 1900 as her year of birth.

If Gentry had become fatigued from the race, her tiredness did not last. The following month, she traveled to North Carolina for a Gentry family reunion and, in September, made her way to Columbus, Ohio, for an OX5 Aviation Pioneers reunion. While there, Gentry was honored with the OX5 Club of America's Distinguished Service Award.

With work, the publicity generated from *Amelia Earhart Lives*, a cross-country air race, a reunion and an award, 1972 was an eventful year.

The years 1973 and 1974 proved to be less active, which allowed Gentry to spend more time on her book. While working at the Belleview-Biltmore Hotel in February 1975, Gentry sent a letter, which is in the collection of the History of Aviation Collection, Special Collections Department in the McDermott Library at the University of Texas at Dallas, to Stella Randolph that said, "My niece Helen Codling will take care of *Hangar Flying* for me, as here at work I do not have the time. A copy will come your way after 2/26/75 hope you like it."

In early 1975, after countless years and diligent work, Gentry's masterpiece had become a reality. Printed and available for purchase, Gentry announced the publication of *Hangar Flying* in letters, one of which is in the collection of the History of Aviation Collection, Special Collections Department in the McDermott Library at the University of Texas at Dallas, that she mailed to friends and acquaintances. The notice read:

> *Dear Friends,*
>
> *Of the many books on early aviation, none is quite like "Hangar Flying." This is an exciting first-person account of the men and women for whom barnstorming, racing, and record flights were a way of life.*
>
> *In "Hangar Flying," you can re-live the excitement of the 1920's and 30's when flying was both hazardous and humorous. Author Viola Gentry has commissioned a limited edition which will be numbered and autographed, at $5.00 postpaid.*
>
> *With over 200 pages and dozens of rare photographs, this Limited Edition will be a rare collector's item.*

The letter closed with Gentry's signature and the address from which to order the book.

Lieutenant General James H. "Jimmy" Doolittle wrote the foreword for *Hangar Flying*. In it, he said:

> *"Hangar Flying" is the interesting and non-technical story of flying in the 20's and 30's.*
>
> *It tells of the barnstorming, the exhibition flying, the racing and the record flights of that period. But most of all it tells of the interesting people who flew.*
>
> *Between World War I and World War II the aviation fraternity was small and almost everyone in it knew almost everyone else. We were therefore a very closely knit group.*

Cover of Viola Gentry's book *Hangar Flying*, 1975. *Courtesy of the author.*

In reading the manuscript of "Hangar Flying" I had the distinct pleasure of visiting—vicariously—with old and dear friends...Many of whom are gone, or as Viola so eloquently puts it, "Have folded their wings."

Today's pilots have to know a great deal more than the pilots of that era but I doubt if they will ever have as much fun...some of us still enjoy the nostalgia of harkening back to the days when the primary flight instrument was the seat of a pilot's pants...

> *I recommend to the old-time pilot, to today's modern pilot, to all of those interested in aviation, this fine book written by a lovely and gracious lady who...still has the indomitable spirit which made her a great pilot and loyal friend.*

Hangar Flying is perhaps Gentry's greatest legacy. Within its pages, Gentry provides a glimpse of her own life and, through personal remembrances and firsthand accounts, reveals the perils of early flight. *Hangar Flying* is a vital record of aviation history, a testament to Gentry's passion for flight and those who flew.

When *Hangar Flying* was released, Gentry was eighty years old and a victim of cataracts. The "spunky" woman who soloed in 1925, who never allowed any circumstance to clip her wings, would not pilot an airplane again. Despite this malady, Gentry continued to promote aviation, preserve its history, attend and arrange aviation-related events and be honored with aviation awards.

On July 24, 1976, Gentry's contributions to aviation were recognized by the International Forest of Friendship in Atchison, Kansas. On that day, Gentry's name became a permanent fixture on the Forest's Memory Lane, a walkway embedded with the names of honorees. The memorial was sponsored by the City of Atchison, the Ninety-Nines and the University of Kansas Forestry Extension. It is unknown if Gentry attended the dedication, but there is no reason to think she did not.

That same year, a committee of the Ninety-Nines began collecting memorabilia and artifacts relating to the history of women in aviation. Its reason for doing so was to establish a museum—the International Women's Air & Space Museum (IWASM).

Doris Scott, former president of the IWASM, sent a letter to Gentry on September 29, 1976. Scott told of the museum's progress and stated:

> *Since I do not know all of you personally and I have heard that many of you have contributed quite a bit to the advancement of aviation, we would like to have documentation, pictures, books, artifacts and memorabilia on you so that we could have it for the museum and also for the first International Aviation Women's Library...*
>
> *If you know of any other women outside our organization that have contributed to the advancement of aviation, I would appreciate you sending me their names and addresses so that they can be contacted immediately.*

Scott's attempt to collect material for the new museum was reminiscent of Gentry's efforts to contact Early Birds on behalf of the HAC at the University of Texas at Austin. It was no surprise, therefore, that Gentry eagerly responded to the letter with information on herself, a copy of *Hangar Flying* and a list of names and organizations Scott should contact. Scott responded to Gentry by saying, "I am so glad to have you helping us because of your long-standing expertise on women in aviation."

As a result of Scott's request and the museum's dedication to the preservation and dissemination of the history of women in aviation, the IWASM became a repository for many Gentry-specific items.

By the latter part of the 1970s, Gentry had retired from her housekeeping positions and was a full-time resident of Miami, Florida. On May 20, 1977, she returned to New York for the fiftieth-anniversary celebration of Charles Lindbergh's nonstop flight from New York to Paris. Gentry knew Lindbergh and spoke of him in 1972. An article written by Irene Albert for the *Clearwater Sun* on January 23, 1972, titled "Aviatrix Is Birdmen's Celebrity" told of Gentry's aerial adventures, as well as her memories of Lindbergh. Albert stated:

> *The enthusiastic aviatrix never expects to live through a more exciting moment than when she saw Lindbergh take off…for his trans-Atlantic solo flight to Paris in 1927.*
>
> *"People stayed up all night to see him leave," she said. "It was far more exciting than watching the astronauts start for the moon…and in those days, just about as dangerous." During several years of association with him, she found Lindbergh "just as shy then as he has always been, but very human and friendly."*

While attending the anniversary festivities, Gentry was interviewed by George Vecsey. In his book *Getting Off the Ground*, Gentry is featured in the chapter titled "Age of Heroes."

Vecsey highlighted Gentry's early flying career and also shed light on her activities in 1977. He said:

> *Now she travels around the country to as many reunions as she can afford—a gentle pioneer who does not care for addressing large groups, but who can charm a small audience. Today, at the Lindbergh ceremony, when Viola Gentry is introduced to the crowd, she waves timidly. But the young newscasters, with their persistent microphones and their limited knowledge of Gentry's history, find her a charming interview.*

Viola Gentry (right) with her sister, Thelma Gentry Hayes, early to mid-1970s. *Courtesy of Helen H. Codling.*

For the remainder of 1977, Gentry continued to participate in aviation events, as time and money permitted. In early 1978, Gentry was overcome with sadness when her beloved sister, Thelma, who was eighty-one, took ill and died. At the time of her death, which took place on January 15, Thelma was a resident of Memphis, Tennessee. She was interred at Leemont Cemetery in Danville, Virginia.

The remaining years of the decade came and went quietly for Gentry. Newspapers did not document her appearance at aviation-related gatherings, and museum files are void of letters or photographs. Yet in 1980, Gentry's name was heralded once again when she was inducted into the OX5 Aviation Pioneers Hall of Fame. The hall of fame, according to the 2004 OX5 Membership Roster, honors "distinguished aviation people who have contributed significantly to the progress, growth or safety of aviation on a national scale."

The OX5 Aviation Pioneers Hall of Fame, located at the Glenn H. Curtiss Museum in Hammondsport, New York, honors more than three hundred individuals. Plaques on the OX5 Aviation Pioneers Hall of Fame provide a photograph of the honoree, as well as a brief biography. Gentry's plaque reads:

> *VIOLA GENTRY*
> *Established first official solo endurance record for women, 1928.*
> *Although crippled in a crash while attempting a refueling record in 1929, she competed later in three Powder Puff Derbies and in an Angel Race.*
> *Charter Member 99's—an Honorary Early Bird and an OX5er.*
> *Author of "Hangar Flying" covering early aviation personalities.*
> *INDUCTED 1980*

Thanks to the hall of fame, Gentry's name and major accomplishments are recorded for posterity. She, like the other honorees, will be remembered as an icon of aviation.

Following her induction into the OX5 Aviation Pioneers Hall of Fame, Gentry slipped from the public's view and into the twilight of her years.

Viola Gentry's OX5 Aviation Pioneers Hall of Fame card, 1980. *Courtesy of Helen H. Codling*

Perhaps, as she approached her nineties, Gentry decided it was time to rest from her travels and to forego aviation-related functions. Maybe her health, particularly her eyesight, prevented her from attending gatherings and doing the things she loved.

On June 23, 1988, at the age of ninety-four, Viola Estelle Gentry folded her wings. The woman who struggled to pay for flying lessons, flew during the dawn of aviation, experienced discrimination, saw the acceptance and progression of female pilots and witnessed aerial feats by the legends of aviation had taken her final flight.

Beyond Gentry's accomplishments in aviation, she inspired all who knew her to never give up and to never forfeit a dream. Despite being plagued with countless challenges, Gentry trudged forward. She never stopped, looked back or turned around. Gentry's life validates the fact that one can never be too poor, or too old, to fulfill a dream.

When Gentry died in Miami, Florida, there were no large gatherings or grand speeches from famous men and women. Nevertheless, no words could have defined Gentry's life any better than two sentences that were printed in the *Danville Register & Bee* on June 25, 1988. The article "Aviatrix with Danville Ties Dies" proclaimed, "She wanted to fly airplanes. And fly she did."

"And fly she did," indeed.

Bibliography

Manuscripts

Gentry, Viola E. *The Baby Cardinal.* N.d. TS. Collection of the International Women's Air & Space Museum, Cleveland, OH.

———. *The Birds' Musical.* N.p.: privately printed, circa 1970. TS. Collection of the International Women's Air & Space Museum, Cleveland, OH.

———. *Chirpy and Croakey.* N.d. TS. Collection of the International Women's Air & Space Museum, Cleveland, OH.

———. *The Patch.* N.d. TS. Collection of the International Women's Air & Space Museum, Cleveland, OH.

U.S. Department of the Treasury. Internal Revenue Services. Viola Gentry Warren's "Application for Social Security Account Number." Washington, D.C.: GPO, 1940.

———. *Statistics of Income from Returns of Net Income for 1924.* Washington, D.C.: GPO, 1926.

Wert, Hal. "Kansas City in WWII: Recovery and Transformation." Kansas City Museum, April 2013. http://www.kansascitymuseum.org/CURATOR/curator_wert2.html.

Newspapers

Albuquerque Tribune (Albuquerque, NM)

Anderson Herald (Anderson, IN)

Austin American (Austin, TX)

Bee (Danville, VA)

Bridgeport Post (Bridgeport, CT)
Bridgeport Telegram (Bridgeport, CT)
Brooklyn Daily Eagle (Brooklyn, NY)
Brooklyn Eagle (Brooklyn, NY)
Brooklyn Standard Union (Brooklyn, NY)
Charlotte Observer (Charlotte, NC)
Chillicothe Constitution-Tribune (Chillicothe, MO)
Clearwater Sun (Clearwater, FL)
Daily News (Huntingdon, PA)
Daily Northwestern (Oshkosh, WI)
Daily Star (Queens, NY)
Danbury Reporter (Danbury, NC)
Danville Register (Danville, VA)
Davenport Democrat and Leader (Davenport, IA)
Decatur Evening Herald (Decatur, IL)
Dunkirk Evening Observer (Dunkirk, NY)
Evening Independent (St. Petersburg, FL)
Evening News (San Jose, CA)
Fort Worth Star-Telegram (Fort Worth, TX)
Geneva Daily Times (Geneva, NY)
Goldsboro News-Argus (Goldsboro, NC)
Greensboro Telegram (Greensboro, NC)
Henry Bulletin (Martinsville, VA)
Iowa City Press-Citizen (Iowa City, IA)
Kane Republican (Kane, PA)
Kansas City Star (Kansas City, MO)
Kingsport Times (Kingsport, TN)
Landmark (Statesville, NC)
Ludington Daily News (Ludington, MI)
Madison Messenger (Madison, NC)
Mansfield News-Journal (Mansfield, OH)
Martinsville Daily Bulletin (Martinsville, VA)
Miami Daily News (Miami, FL)
Miami Herald (Miami, FL)
Middletown Times Herald (Middletown, NY)
Milwaukee Sentinel (Milwaukee, WI)
Newark Evening News (Newark, NJ)
News & Observer (Raleigh, NC)
New York Sun (New York City, NY)
New York Times (New York City, NY)
Ocean County Review (Seaside Heights, NJ)

Ogden Standard-Examiner (Ogden, UT)
Pampa Daily News (Pampa, TX)
Post-Crescent (Appleton, WI)
Poughkeepsie Eagle-News (Poughkeepsie, NY)
Prescott Evening Courier (Prescott, AZ)
Redwood City Tribune (Redwood City, CA)
Reidsville Review (Reidsville, NC)
Reno Evening Gazette (Reno, NV)
Roanoke Times (Roanoke, VA)
San Francisco Chronicle (San Francisco, CA)
San Jose Mercury (San Jose, CA)
Schenectady Gazette (Schenectady, NY)
Scranton Republican (Scranton, PA)
Sheboygan Press (Sheboygan, WI)
Statesville Record & Landmark (Statesville, NC)
Syracuse Herald (Syracuse, NY)
Times (Hammond, IN)
Tucson Daily Citizen (Tucson, AZ)
Utica Daily Press (Utica, NY)
Warrensburg–Lake George News (Warrensburg, NY)
Washington Post (Washington, D.C.)
Wellsboro Gazette (Wellsboro, PA)
Wichita Beacon (Wichita, KS)
Yonkers Statesman (Yonkers, NY)

Books

Bohrer, Walt, and Ann Bohrer. *Tales Up!* Fallbrook, CA: Aero Publishing, 1971.

Brooks-Pazmany, Kathleen. *Women in Aviation, 1919–1929*. Washington, D.C.: Smithsonian Institution Press, 1991.

Gentry, Viola. *Hangar Flying*. Chelmsford, MA: Privately published, 1975.

Holden, Henry M., and Captain Lori Griffith. *Ladybirds: The Untold Story of Women Pilots in America*. Mount Freedom, NJ: Black Hawk Publishing Company, 1991.

Kirschbaum, Dick. *Fifty Famous Flyers*. Newark, NJ: Charles Dewar Company, 1941.

Klaas, Joe. *Amelia Earhart Lives*. New York: McGraw Hill Book Company, 1970.

Oakes, Claudia M. *United States Women in Aviation, 1930–1939*. Washington, D.C.: Smithsonian Institution Press, 1985.

Parramore, Thomas C. *First to Fly: North Carolina and the Beginnings of Aviation*. Chapel Hill: University of North Carolina Press, 2002.

Planck, Charles E. *Women with Wings*. New York: Harper and Brothers Publishers, 1942.

Plehinger, Russell. *Marathon Flyers*. Detroit, MI: Harlo Press, 1989.

Scott, Catherine D., ed. *Aeronautics and Space Flight Collections*. New York: Haworth Press, Incorporated, 1985.

Vecsey, George, and George C. Dade. *Getting Off the Ground*. New York: E.P. Dutton, 1979.

Welch, Roseanne. *Encyclopedia of Women in Aviation and Space*. Santa Barbara, CA: ABC-CLIO, Incorporated, 1998.

Articles

Buffington, H. Glenn. "Flying Life of Viola Gentry." *Journal of the American Aviation Historical Society* 13, no. 2 (Summer 1968): 125–26.

The Ninety-Nines, Inc. News Letter, August 1960 and July/August 1961.

the99News, August/September 1972.

Parramore, Tom. "Viola Gentry." *Coastland Times*, October 30, 1997.

Pauley, Robert F. "The Paramount Aircraft Corporation." *Skyways*, July 2001.

Internet Sources

Ancestry.com. Birth, Marriage, Divorce and Death Records. www.ancestry.com (accessed 2012–14).

———. Draft Enlistment and Service Records. www.ancestry.com (accessed 2012–14).

———. U.S. City Directories. www.ancestry.com (accessed 2012–14).

———. U.S. Federal Census. www.ancestry.com (accessed 2012–14).

Florida Memory. Historic images and vital statistics. www.floridamemory.com (accessed 2012–14).

Library of Congress. Prints and Photographs Online Catalog. www.loc.gov/pictures (accessed 2012–14).

Newspapers.com. Historical newspapers. www.newspapers.com (accessed 2012–14).

Old Fulton New York Postcards. Historical newspapers. www.fultonhistory.com (accessed 2012–14).

Index

A

Adams, Bernetta 162
Adams, Clara 115, 170
Adelphi College Center for Long Island Women in Home Defense 123
Air Commerce Act of 1926 42
Aircraft Accessories Corporation 128, 129, 130, 131
Aircrafters of Philadelphia 114
airlines
  American 170
  Australia 152
  Eastern 132
  Pan American World Airways 152, 153, 158
airplanes
  Arrow Sport 67, 68, 70
  Atlantic-Fokker C-2A 71
  Bellanca 72
  Bellanca Super Viking 171
  Boeing 747 170
  Curtiss JN-4 34
  Curtiss Oriole 35
  Curtiss Thrush 105, 110
  Golden Eagle Monoplane 58
  Lockheed Vega 98
  Monocoupe 70 102
  Paramount Cabinaire 70, 71, 73, 78
  Piper Comanche 170
  Piper PA-22 Tri-Pacer 148, 149, 150
  Rearwin Skyranger 123
  Ryan B-1 Brougham 73
  Ryan Navion B 144, 145, 147
  Swallow 62, 66, 70
  Travel Air 46, 48
  Waco 109
airports
  Amelia Earhart 152
  Holmes 64, 65, 66, 76, 78, 85
  Torrance Municipal 147
Alexander, Mary 91
All-American Air Races 101
All-Woman Transcontinental Air Race 144, 145, 146, 148, 150, 168, 171
Amelia Earhart Domestic Airmail Stamp 151, 152, 153
Amelia Earhart Scholarship Fund 131, 151, 152
American Red Cross 24, 120
Annette Gipson All-Women Air Race 107, 109
Answer, The 71, 72, 74, 75, 79, 81, 82, 83, 86, 87, 88

Ashcraft, Francis 76
Ashcraft, John W. "Big Jack" 75, 76, 77, 79, 81, 86, 129, 146

B

Bastie, Maryse 53
Bayley, Arnold Blakeman 137, 143, 144, 145, 146, 147, 148, 149, 157, 170
Beacon Manor Hotel 171
Beech, Olive Ann 152
Bellanca, Dorothy (Mrs. Giuseppe) 163
Bellanca, Giuseppe M. 163
Belleview-Biltmore Hotel 135, 136, 141, 143, 173
Bennett, Cora (Mrs. Floyd) 110, 111, 154
Bennett, Floyd 154
Benoist, Thomas 163
Bergin, Emil 74
Beta Sigma Phi 158, 160
Birdman Trophy 109
Bishop, Wallace "Wally" 85, 96
Black, Gretchen S. 158
Bohrer, Walter "Walt" 162
Bolam, Guy 167
Bolam, Irene O'Crowley Craigmile 166, 167
Booth, Harry 49
Brăescu, Smaranda 101
Bridgeport District Ordnance Department 23, 24
bridges
  Brooklyn 35, 36, 37, 38, 39, 40, 41
  Manhattan 36, 37, 38, 39, 41
Brimm, Daniel J. 93
British Air Transport Auxiliary 126
Broadwick, Tiny 155, 156, 163
Brooks, William C. "Whispering Bill" 78
Brown, Margery 91
Brown, Mayor George A. 60
Brown, Rives S. 57, 60
Budwig, Gilbert G. 101
Buranelli, Felicity 157, 158, 159
Burnelli, Vincent 157, 158

C

Cagle, Myrtle "Kay" Thompson 144, 145, 146, 147, 148
California
  San Carlos 170
  San Diego 148
  San Francisco 24, 25, 26, 27, 28, 29, 32
  San Jose 162
  Santa Monica 28, 29
  Torrance 145, 147
Cameron, Alexander McNight 130, 131, 132, 135
Campbell, Louise Hayes 131
Caperton, Arthur "Art" 36
Carr, Walter J. 70, 72, 73
Chamberlin, Clarence 65
Chamberlin, Jessie R. 118
Chassey, Irene 91
Civil Aeronautics Administration 127
Civil Aeronautics Authority 117
Clarke, H.B. 66, 72
Cleveland Air Races 113
Cochran, Jacqueline "Jackie" 126
Codling, Helen Hayes 19, 24, 103, 115, 129, 130, 140, 173
Connecticut
  Bridgeport 22, 23, 24, 86
  Hartford 78, 82
Craigmile, Charles 167
Croninger, Martha Helen 44
Crosson, Marvel 29
Cryer, Mayor George E. 44
Curtiss Factory 118
Curtiss, Glenn H. 158, 163

D

DeBever, Charles 95, 96
Delaware
  Wilmington 145, 147
Department of Commerce 42, 43, 44, 50, 94, 100, 104, 131
Descomb, Edith 101
Desert Honey 149
Dixon, Carl 82

Doolittle, James H. "Jimmy" 46, 173
Drummond-Hay, Lady Grace Marguerite Hay 130, 138

E

Eanes, Minnie 24
Earhart, Amelia M. 41, 45, 46, 66, 91, 104, 105, 115, 124, 131, 132, 151, 152, 158, 166, 167, 172
Earhart, Amy Otis 131
Elder, Ruth 44
Enslow, Randy 96

F

Federal Aviation Administration 43, 95, 101, 127, 131, 135, 169
Fédération Aéronautique Internationale 34, 35, 42, 51, 107
fields
  Cicero 118, 119
  Clover 29
  Crissy 27, 28, 29
  Curtiss 31, 33, 34, 35, 36, 42, 43, 46, 49, 52, 64, 67, 73, 80, 91, 109, 129, 141, 146
  Floyd Bennett 99, 100, 102, 107, 110, 111, 114, 115, 127, 131, 136, 138
  Kennedy 158
  LaGuardia 130
  Meacham 73
  Roosevelt 44, 46, 47, 49, 50, 52, 56, 61, 62, 66, 70, 72, 73, 74, 77, 78, 79, 82, 85, 86, 89, 93, 95, 98, 102, 103, 113, 123, 129, 141, 167
First Flights 115
Fitzgerald, Helen G. 101
Florida
  Belleair 135, 136
  Dade County 135
  Jacksonville 20, 21
  Miami 101, 111, 115, 124, 125, 126, 160, 163, 167, 171, 176, 179
  Tampa 131
Florida Ostrich Farm 21, 88
Flying Boudoir, The 105, 106, 111
Fonck, René 41, 62
Fort Worth 73
Foster, Benjamin 35
Fowler, Robert "Bob" 27, 28, 163
Frank, K.G. 52
Fredeman, Mrs. Frank H. 44

G

Gates Flying Circus 75, 77, 85
Gates Flying Service 75, 76
Gatty, Harold 98, 99, 100, 102, 104
Gee, Elizabeth 20
Gee, George Henry 20, 22, 24
Gee, Joshua 20
Gentry, Anna 17, 18, 19
Gentry, Benson 17
Gentry, Clyde 17
Gentry, James Forrest 19
Gentry, Lynwood Inez 19
Gentry, Maydie Blanche Price 18, 19, 39, 54, 133
Gentry, Nettie Walters 15, 17
Gentry, Richard J. 17, 18
Gentry, Samuel Garrett 17, 18, 19, 20, 39, 54, 59, 83, 133
Gentry, Thelma 15, 18, 19, 83
Gentry, Walter 17
Georgia
  Macon 75, 76, 77
Gervais, Joe 166, 167
Gillis, Fay 91
Gilmore, Lyman 163
Gipson, Annette 108, 124
Glenn H. Curtiss Museum 178
Glenn, John H., Jr. 157, 158
Goldschmidt, Martha E. 94
Goodrich, Mary 91
Grand Hotel, California 24, 25, 27
Grand Hotel, Michigan 169
Graveley, Spot 20
Gray, George A. 21, 22, 86, 88
Gray, "Jack" Stearns 88

Grey, Geraldine 44
Griese, Jane 146, 148
Griffin, Gloria Conti 141

H

Haddaway, George 157, 161, 165
Haldeman, George 65, 163
Harrell, Frances 91
Harvey, Al 19
Harvey, Claud 19
Harvey, Ed 156
Harwood, Oren P. 104
Havens, Beckwith 163
Hawaii
  Honolulu 132
Hawks, Frank M. 158
Hayes, Oakley Cabell 44, 131
Hayes, Thelma Gentry 44, 54, 82, 83, 86, 131, 177
Heath, Lady Mary Sophie 53
Hennicke, Carl A. "Slim" 166
Hicks, Edwin 82
Hicks, Henry 82
Hicks Nursery 81
History of Aviation Collection at the University of Texas at Austin 160, 161, 162, 163, 165, 173, 176
Holmes, E.H. 64, 65
Hoover, Herbert 154
Hopper, Anna Gentry 18
Hopper, George W. 18
Hopper, Richard Gentry 19
Hospital for the Ruptured and Crippled 89, 91, 95, 97
Hotel Biltmore 136
Hotel New Yorker 119, 121
Hotel Webster 141, 143
Hoyt, Jean David 91
Hubbard, Bernard R. 158
Hurst, Fanny 115
Huyler, Betty 91

I

I.J. Fox Fur Company 105
International Feminine Endurance Record 53
International Forest of Friendship 175
International Women's Air & Space Museum 46, 49, 88, 99, 102, 117, 118, 130, 137, 160, 165, 175
I Wanted Wings 123

J

James, Joseph R. 73
Jensen, Marguerite "Peg" 79, 83
Jensen, Martin 72, 79, 83
Johnson, Ruth 170, 171
Jones, Charles S. "Casey" 31, 104, 105, 106

K

Kansas
  Atchison 151, 152, 175
  Protection 75, 86
  Wichita 62
Keith-Miller, Jessie 91
Kelly, James 73
Kenyon, Cecil "Teddy" 91
Kincaid, Claude 82
Kirschbaum, Dick 115, 122
Korn, Ed 163
Korn, Milton 163
Krantz, Duke "Diavalo" 75
Kunz, Opal Logan 91

L

Lady Hay Drummond-Hay Memorial Trophy 137, 138, 139, 140
Lahm, Frank P. 158
Leh, Dorothea 91
Lehigh, Rod 122
Lesser, Marjorie May 91
Lindbergh, Charles 41, 109, 176
Locklear, Ormer L. 25, 26, 27
Logsdon, C.S. 136, 137
Long Island Early Fliers Club 144, 147, 149, 166, 167
Louisiana
  Baton Rouge 76

Lund, Bettie (Mrs. Freddie) 101, 102
Lyon, Grace 46, 48, 50, 89

M

MacCracken, William P., Jr. 42
Madame Barna's French Restaurant 33
Marsalis, Frances Harrell 91, 105, 106, 111, 112, 113
Marshall, Shirley 146, 148, 149, 150, 151, 152, 153
Massachusetts
 Worcester 67, 68, 69, 70
Massachusetts State Insane Hospital 68, 69
Masson, Angela 170, 171
Mathews, Keet 91
Maynard, Belvin W. 40
McCory, Herbert 74
McKenzie, John 99
McKinney, Ive 78
McWilliams, Irwin Keyes 68, 69
Medal-of-the-Month Club 157
Merrill, Mazel M. 31
Michigan
 Mackinac Island 169
Miller, Lestere 163
Missouri
 Jackson County 130
 Kansas City 127, 128, 129, 130, 131
Mitchell, William Lendrum 158
Monmouth Hotel 172
Moore, Mrs. Robert 101
Morton, Hugh 67
Musick, Edwin C. 158

N

Nassau County Hospital 84, 86, 89, 90, 91
National Aeronautic Association of the United States 34, 35, 42, 49, 51, 53, 107, 136
National Aeronautics and Space Administration 153, 154
National Recovery Administration 109
Nelson, Sylvia A. 91
New Jersey
 Atlantic City 48, 148
 Belleville 103
 Orange 114
 Point Pleasant Beach 171
 Spring Lake 172
 Toms River 170
New York
 Brooklyn 33, 35, 36, 37, 38, 39, 40, 41, 99, 100, 103, 119, 127, 131
 Coney Island 109
 East Hampton 137, 138, 145, 147, 152, 154
 Hammondsport 178
 Hempstead Plains 62
 Jackson Heights 64, 76, 78
 Long Beach 46
 Long Island 31, 44, 46, 49, 62, 64, 67, 72, 76, 80, 81, 82, 84, 91, 118, 123, 129, 137, 138, 146, 162
 Manhattan 36, 37, 38, 39, 41, 89, 122, 132
 Middletown 96
 Mineola 44, 84
 New York City 96, 109, 121, 122, 141
 Valley Stream 31, 91
 Warrensburg 154
New York Women's Press Club 148
Nichols, Ruth 44
Ninety-Nines 91, 92, 93, 126, 131, 132, 140, 144, 147, 148, 149, 151, 157, 158, 175
Nolen, Pat 151, 152, 153
Noonan, Fred 115
North Carolina
 Gentry 15, 17
 Goldsboro 158, 159
 Greensboro 19
 Rockingham County 15, 17, 19, 49
 Selma 144, 146
North Miami Woman's Club 124

O

Ober, Willy 163
Ohio
  Columbus 172
O'Mara, Margaret F. 91
Omlie, Phoebe 44
Operation Earhart 166, 167
Outdoor Girl 110, 111, 112
Outdoor Girl Cosmetics 110
OX5 Aviation Pioneers 34, 155, 156, 172, 177, 178
OX5 Aviation Pioneers Hall of Fame 177, 178
OX5 Club of America 155, 156, 164, 172

P

Pangborn, Clyde 74, 75, 78
Paramount Pictures, Incorporated 123
Paris, Neva 91
Parkhurst, Charles W. 73, 74
Pennsylvania
  Philadelphia 31, 32, 33, 114, 123
  Towanda 75, 77
Pickerill, Elmo N. 128, 129, 141, 143, 146, 147, 148, 163, 166
Post, Mrs. Wiley 98
Post, Wiley 98, 99, 100, 102, 104
Powder Puff Derby 144, 147, 148
Powell, Captain 153
Price, James R. 19
Purdue University 151

Q

Question Mark 71, 72, 73
Quiet Birdmen of New York 114

R

Randolph, Stella 163, 173
Rawlings, L.B. 64
Remey, Mrs. John T. 101
Rhode Island
  Providence 22
  Warwick 77
Richardson, D.G. 43
Richey, Helen 112, 113
Robbins, Reginald "Reg" 73
Robinson Circus 19
Rockwell, Lee 70, 89
Rodgers, Cal 163
Roger Q. Williams School of Aeronautics 33
Roosevelt Field Inn 120
Roosevelt, Franklin D. 109
Roosevelt Hotel 122, 123
Rotholz, Meta 91
Royal Order of Flying Jackasses 64

S

Salvation Army 96, 97
Sansom, Mary 110, 111
Schneider, Carl 53
Scott, Blanche Stuart 163
Scott, Doris 175
Sears, John 22, 23, 24
Sears, Mrs. John 22, 23, 24, 86
Sea Spray Inn 137, 138, 143, 144, 145, 147, 148, 156, 166
Sesquicentennial International Exposition 31, 32, 33, 123
Silver Wings Fraternity 144
Smith, Elinor 41, 62, 63, 66
Smith, Lowell 29
Smith, Sylvia 98
Southey, Earl 109
Spirit of Ammonia 77
Steele, Samuel "Buck" 76
St. Francis Hotel 25, 26, 27
Stiles, Jean Lenore 101
Stinson, Marjorie 44
Stone, Dorothy 65

T

Tennessee
  Memphis 177
Texas
  Austin 157
  Fort Worth 73, 164
  Temple 157

Thaden, Louise McPhetridge 44, 45, 105, 106
3 Musketeers 72, 79, 83, 84
Tiny Broadwick Award 155, 156
Trout, Evelyn "Bobbi" 57, 58, 59

U

Uhl, Betty Robertson 155
Ulbrich, William "Bill" 50, 52, 61, 62, 67, 68, 69, 72, 79, 83, 84, 86
United States Army Air Corps 71, 122, 123
United States Civil Service Commission 117, 118
University of Texas at Austin 161, 162, 164, 165, 176

V

Van Bechten, Katherine Schuyler 44
Viola Gentry Skyline Fashions 106
Virginia
 Danville 18, 19, 20, 22, 24, 39, 54, 56, 59, 82, 133, 137, 138, 177, 179
 Martinsville 44, 45, 55, 56, 57, 58, 59, 60, 61, 72, 82, 86
 South Boston 61

W

Wade, Leigh 29
Waite, H. Roy 163
Walker, James J. 41, 104
Walsh, Wilma L. 91
Warner, Elizabeth 44
Warren, John R. "Jack" 102, 103, 113, 114, 115, 132
Warren, Willie 20
Washington
 Spokane 77
Webb, E. Ruth 91
Webster, Clifford L. 163
Wells, Otis 20
Westhampton Air Force Base Officer's Club 144, 147, 166
Whitehead, Gustave 163
Wilburn, Mayor 61
Wild, Horace B. 117, 118, 119, 163
Williams, Roger Q. 33, 86, 110
Winnie Mae 98
Women Airforce Service Pilots 144, 170
Women's Air Derby 105, 144
Women's Emergency Aid Committee for Unemployment 96
Women's International Association of Aeronautics 114, 117, 123, 127, 131, 138, 139
Woodruff, Harold 157
World War I 20, 22, 24, 25, 75, 109, 123, 162, 173
World War II 123, 126, 127, 131, 144, 173
Wright, Orville 15, 17, 19, 157, 158, 163
Wright School of Aviation 21
Wright, Wilbur 15, 17, 19, 157, 158, 163

Y

Yancey, Lewis A. 86
Young, Clarence M. 43, 94, 101

About the Author

Jennifer Bean Bower is an award-winning author, native Tar Heel, lifelong resident of Winston-Salem and graduate of the University of North Carolina at Greensboro. A passionate student of North Carolina history, Bower seeks to document the lesser-known people, places and events of her state's past.

While researching Maynard Field, North Carolina's first commercial airfield, Bower was introduced to Viola Estelle Gentry. So inspired by Gentry's determination to fulfill her dream of flight, Bower began researching and documenting the pilot's life and aviation endeavors. As Gentry's life began to take shape from the words of historical newspapers and documents, Bower was compelled to share her story. In addition to the biography, Bower has created a first-person narrative that highlights the triumphs and tragedies of Viola Gentry's life.

Bower is the author of numerous articles, all of which relate to North Carolina history, and three other books: *Animal Adventures in North Carolina*; *Winston & Salem: Tales of Murder, Mystery and Mayhem*; and *Moravians in North Carolina*. Should you wish to contact Bower or stay up to date on her current projects, please visit www.JenniferBeanBower.com.